AF899449

Practical Mathematics in a Commercial Metropolis

Archimedes

NEW STUDIES IN THE HISTORY AND PHILOSOPHY OF SCIENCE AND TECHNOLOGY

VOLUME 31

EDITOR

JED Z. BUCHWALD, *Dreyfuss Professor of History, California Institute of Technology, Pasadena, CA, USA.*

ASSOCIATE EDITORS FOR MATHEMATICS AND PHYSICAL SCIENCES

JEREMY GRAY, *The Faculty of Mathematics and Computing, The Open University, Buckinghamshire, UK.*
TILMAN SAUER, *California Institute of Technology*

ASSOCIATE EDITORS FOR BIOLOGICAL SCIENCES

SHARON KINGSLAND, *Department of History of Science and Technology, Johns Hopkins University, Baltimore, MD, USA.*
MANFRED LAUBICHLER, *Arizona State University*

ADVISORY BOARD FOR MATHEMATICS, PHYSICAL SCIENCES AND TECHNOLOGY

HENK BOS, *University of Utrecht*
MORDECHAI FEINGOLD, *California Institute of Technology*
ALLAN D. FRANKLIN, *University of Colorado at Boulder*
KOSTAS GAVROGLU, *National Technical University of Athens*
PAUL HOYNINGEN-HUENE, *Leibniz University in Hannover*
TREVOR LEVERE, *University of Toronto*
JESPER LÜTZEN, *Copenhagen University*
WILLIAM NEWMAN, *Indiana University, Bloomington*
LAWRENCE PRINCIPE, *The Johns Hopkins University*
JÜRGEN RENN, *Max-Planck-Institut für Wissenschaftsgeschichte*
ALEX ROLAND, *Duke University*
ALAN SHAPIRO, *University of Minnesota*
NOEL SWERDLOW, *California Institute of Technology, USA*

ADVISORY BOARD FOR BIOLOGY

MICHAEL DIETRICH, *Dartmouth College, USA*
MICHEL MORANGE, *Centre Cavaillès, Ecole Normale Supérieure, Paris*
HANS-JÖRG RHEINBERGER, *Max Planck Institute for the History of Science, Berlin*
NANCY SIRAISI, *Hunter College of the City University of New York*

Archimedes has three fundamental goals; to further the integration of the histories of science and technology with one another: to investigate the technical, social and practical histories of specific developments in science and technology; and finally, where possible and desirable, to bring the histories of science and technology into closer contact with the philosophy of science. To these ends, each volume will have its own theme and title and will be planned by one or more members of the Advisory Board in consultation with the editor. Although the volumes have specific themes, the series itself will not be limited to one or even to a few particular areas. Its subjects include any of the sciences, ranging from biology through physics, all aspects of technology, broadly construed, as well as historically-engaged philosophy of science or technology. Taken as a whole, *Archimedes* will be of interest to historians, philosophers, and scientists, as well as to those in business and industry who seek to understand how science and industry have come to be so strongly linked.

For further volumes:
http://www.springer.com/series/5644

Ad Meskens

Practical Mathematics in a Commercial Metropolis

Mathematical Life in Late 16th Century Antwerp

Drawings by Paul Tytgat

Springer

Ad Meskens
Department Bedrijfskunde, lerarenopleiding en Sociaal Werk
Artesis Hogeschool Antwerpen
Antwerpen
Belgium

ISSN 1385-0180
ISBN 978-94-007-5720-2 ISBN 978-94-007-5721-9 (eBook)
DOI 10.1007/978-94-007-5721-9
Springer Dordrecht Heidelberg New York London

Library of Congress Control Number: 2013933140

© Springer Science+Business Media Dordrecht 2013
This work is subject to copyright. All rights are reserved by the Publisher, whether the whole or part of the material is concerned, specifically the rights of translation, reprinting, reuse of illustrations, recitation, broadcasting, reproduction on microfilms or in any other physical way, and transmission or information storage and retrieval, electronic adaptation, computer software, or by similar or dissimilar methodology now known or hereafter developed. Exempted from this legal reservation are brief excerpts in connection with reviews or scholarly analysis or material supplied specifically for the purpose of being entered and executed on a computer system, for exclusive use by the purchaser of the work. Duplication of this publication or parts thereof is permitted only under the provisions of the Copyright Law of the Publisher's location, in its current version, and permission for use must always be obtained from Springer. Permissions for use may be obtained through RightsLink at the Copyright Clearance Center. Violations are liable to prosecution under the respective Copyright Law.
The use of general descriptive names, registered names, trademarks, service marks, etc. in this publication does not imply, even in the absence of a specific statement, that such names are exempt from the relevant protective laws and regulations and therefore free for general use.
While the advice and information in this book are believed to be true and accurate at the date of publication, neither the authors nor the editors nor the publisher can accept any legal responsibility for any errors or omissions that may be made. The publisher makes no warranty, express or implied, with respect to the material contained herein.

Printed on acid-free paper

Springer is part of Springer Science+Business Media (www.springer.com)

To Guido, Fernand, Chris Jean Paul, Hugo
and of course Nicole, Anke, Eva

Abbreviations Used

ABML	Archives de la Province belge méridionale et du Luxembourg, Brussels
AFL	Archief della Faille – de Leverghem
ARA	Algemeen Rijksarchief, Brussels
EHC	Erfgoedbibliotheek Hendrik Conscience, Antwerp
KB	Koninklijke Bibliotheek, Brussels
KBDH	Koninklijke Bibliotheek, Den Haag
KIK (KIK-IRPA)	Koninklijk Instituut voor het Kunstpatrimonium, Brussels
MPM	Museum Plantin Moretus, Antwerp
SAA	Stadsarchief Antwerpen
SPk	Stedelijk Prentenkabinet, Antwerp

Chapter 4 incorporates material published in Meskens (1996a).
Chapter 6 incorporates material published in Meskens (1994a).
The section on the sector in Chap. 7 incorporates material published in Meskens (1997a).
Chapter 8 incorporates material published in Meskens (1992).

Up to 1576 the so-called Easter style was used to indicate years. A year would start at Easter. Thus 1 January 1528 is 1 January 1529 in our calendar. This is indicated by the letter o.s. (= Old Style).

The currency used in Antwerp was the Carolus guilder (gl.).
1 gl. = 20 stivers (st. or β) = 240 pennies (ϑ)
Other currencies in use, or which are referred to in arithmetic books, were the Flemish pound, the Brabant pound and the Arthois pound.
1 Flemish pound = 20 shilling = 240 groats = 6 gl.
1 Brabant pound = 4 gl.; 1 Arthois pound = 1 gl.

Insets on pages 69, 70, 73, 75, 87, 87, 88 and symbols of cossic numbers are from the books of Valentin Mennher in Erfgoedbibliotheek Hendrik Conscience, Antwerp.

Detail of Philip Galle after Pieter I Bruegel, *De matigheid* (*Temperance*), 1559 from the series *The World of Seven Virtues*. Museum Plantin-Moretus/Prentenkabinet, Antwerpen - UNESCO Werelderfgoed.

Preface

The sixteenth century marks a watershed in the history of science as well as that of the Low Countries. The Copernican system was gaining ground against the backdrop of the religious wars that ravaged Europe.

The High Middle Ages saw the recovery of long distance international trade. The letter of exchange, whereby a merchant could leave his money with a banker in say Venice and, by means of a single piece of paper, could subsequently recuperate it in say Antwerp, made such business much safer. Merchants no longer had to worry about the possibility of being ambushed by highwaymen. The mathematics required for the valuation and conversion of currencies would ultimately lead to the resolution of problems involving quadratic and cubic equations.

Antwerp was a book production centre of the first order, especially with the Plantin empire, which stretched to Paris in the south and Leiden in the north. Printing made it possible to produce relatively large runs of exact copies at a relatively low cost. Moreover, if demand so required, it was quite easy to produce a second edition. This meant that written materials could be distributed much more widely, which in turn allowed more people to take note of science and its teachings. Printing was the primary motor behind the (at least partial) standardization of mathematical symbolism, which emerged in the course of the sixteenth century. It also put an end to trade secrets: the centuries-old tradition whereby skills and know-how were transferred exclusively from master to apprentice was broken; knowledge became a common good.

During the sixteenth century, despite – and in some cases by the grace of – the many wars, a favourable climate developed for mathematics and science to flourish. This was the century in which the definitive mathematization of nature began.

The image of mathematics that emerged during the sixteenth century was one of the science of instruments: astrolabes, quadrants, etc. These are not scientific instruments as we know them today, but rather computing devices and aid to solve practical problems.

Unwittingly, the Church may have brought about the biggest revolution of all: the heliocentric theory. As the calendar and astronomical phenomena were out of pace, the Church proposed a calendar reform. This was important to Christian liturgy, because the date of Easter depends on the beginning of the astronomical Spring. This in turn brought about a dynamism that made some take a closer look at the Ptolemaic geocentric theory.

Undoubtedly, the sixteenth century is *the* age of discovery. The Portuguese and Spanish – soon followed by the French, English and eventually the Dutch – were opening vast new lands in the Americas. Their voyages posed new problems: how to determine one's position at sea. A pressing need was felt for accurate maps, suitable for navigation, and a method for determining longitude. Mercator provided a solution to the first problem when he published his famous world map *Ad Usum Navigantium* (1569). The second problem would haunt mathematicians and seamen alike until well into the eighteenth century.

Which of these changes and developments can be observed in the commercial metropolis that Antwerp was in the last part of the sixteenth century? That is one of the central topics of this book. Did mathematics teachers include the new algebra in their books and pass this knowledge on to their pupils? Did mathematics teachers incorporate the solution of third degree equations in their curricula? Did they teach trigonometry and, if they did, how? Were the changes to the prevailing worldview really as momentous as we generally believe them to have been? How did social changes influence mathematicians? If Antwerp really was an important centre for the practice of mathematics, then why is it not considered to rank among places such as Florence and Amsterdam? This book traces the lives of various mathematicians and their work. Two stand out: Michiel Coignet and, to a lesser extent, Peter Heyns. Starting out as a school teacher, Coignet's career spans almost the entire gamut of practical mathematics. His work covers everything from third degree equations to the Tychonic world system, and hence it is a focal point of this study. His counterpart, Peter Heyns, well versed in mathematics and rhetoric, became known as a champion of the Protestant cause. Their lives are literally interwoven in this story of mathematics in sixteenth-century Antwerp.

This book has been 20 odd years in the making. Many people have travelled with me along the way; some have been companions for the whole journey, others for part of it. I have contracted debts of gratitude I can hardly contemplate being able to repay adequately. Nor is it possible for me to mention all those I am indebted to by name, however grateful I am to all. Nevertheless, some people stand out and those I would like to thank explicitly.

Jean Paul van Bendegem and Hugo Soly read the very first texts and have remained friends.

Guido Pardon was instrumental in making my studies on the history of mathematics possible, by creating career opportunities for me.

The late Jacques Van Damme shared his extensive knowledge on scientific instruments very graciously and without any reservation. The scholarly dinners he organized will always be dear to me.

Many thanks are also due to Christopher Johnstone, a true Scotsman, ever optimistic, with a great heart for others. Chris received me into his home as an exchange student on a language course in Edinburgh many years ago. We have remained friends ever since. He corrected the very first article I wrote in English and has always encouraged me to aim higher.

As anyone witnessing the figure of an astrolabe (p. 184) will agree, the long-standing cooperation with Paul Tytgat (Artesis University College, Antwerp) has proved as invaluable as ever.

Thanks are also due to Dirk Anthierens, Jim Bennett, Henk Bos, Fernand De Nys, Marjolein Kool, Victor Rasquin, Marc Tourweé, Jan van Maanen, Denise Verrept, the della Faille family and my teachers at the Rijkslagere school and Koninklijk Atheneum, Dendermonde.

I am also indebted to Artesis University College for its support in the form of a scientific project grant. The present study would never have materialized without the cooperation of various libraries, particularly the library of the Teacher Training College of the Artesis University College Antwerp, Erfgoedbibliotheek Hendrik Conscience (Antwerp), the Museum Plantin Moretus (Antwerp), Koninklijk Museum voor Schone Kunsten (Antwerp) and the Antwerp City Archives. Stephen Windross read and corrected the English; any remaining errors due to late editing are my own.

The following institutions gave their permission to publish photographs of objects in their possession: the Adler Panetarium (Chicago), Bibliothèque nationale (Paris), KBC Bank NV – Rockoxhuis (Antwerp), Koninklijk Instituut voor het Kunstpatrimonium (KIK-IRPA, Brussels), Koninklijke Bibliotheek Albert I (Brussels), Koninklijke Bibliotheek (Den Haag), Museum Boerhaave (Leiden), Museum of the History of Science (Oxford), Rijksmuseum (Amsterdam), Rijksprentenkabinet (Amsterdam), Séminaire de Tournai, Tesseract (Hastings-on-Hudson), University of Wrocław.

Nicole has stood by my side for more than 25 years. She has given me two dear children, Anke and Eva. None of my texts would have materialized without her support.

Contents

Chapter 1
Introduction

1.1 The Low Countries in the Sixteenth Century[1]

In 1555, the ageing Emperor Charles V (1500–1558), leaning on the shoulder of the young William, Prince of Orange (1533–1584), read out his abdication speech, proclaiming Philip (1527–1598) as the new ruler of the Seventeen Provinces. Under Charles V, the Hapsburg Netherlands had gained recognition as a single and separate entity, sovereignty over which would pass to his heirs.

Although Charles V was now the most powerful ruler of Europe, he faced the threat of a growing Protestantism, which led him to establish *an* Inquisition in the Low Countries. Local governments, for their part, tended to be lenient towards Protestantism, fearing disturbances of trade with, among others, the German towns. The Netherlands Inquisition was aware how widespread Protestantism was and therefore targeted the cultural and intellectual elite: printers, schoolmasters and officials. Executions were few and far between, but they did work as a deterrent.

Economically the Low Countries were flourishing: circumstances had been very favourable since the end of the fifteenth century. Flanders and Brabant were very much the motor of this economic boom, generating about 60% of tax revenues, while Holland was slowly gaining in importance. The years 1540–1565 saw an acceleration of the economic expansion. Despite, or perhaps even because of this prosperity, noblemen and merchants alike were less and less inclined to sacrifice their interests for those of a monarchy that strove for global hegemony.

Philip was not the polyglot his father was, and his uncompromising Spanish Catholic education made him a bigot in the eyes of the locals. Nevertheless, his victory over the French at Saint-Quentin forced the latter into a peace treaty (Cateau-Cambrésis, 1559), promising the Low Countries peace in their time. This led Philip, despite warnings from his advisers, to return to Spain.

---

[1]Based on: de Groof (1995), de Jong (2005), Israel (1989, 1998), Janssens (1976, 1981, 2009), Parker (1972, 1990), Roosens (2005), Schoffer et al. (1988), Soen (2007), Swart (1984), Thomas (2004c), van Eysinga (1959), Vermeir (2011), and Woltjer (1994).

A. Meskens, *Practical Mathematics in a Commercial Metropolis: Mathematical Life in Late 16th Century Antwerp*, Archimedes 31, DOI 10.1007/978-94-007-5721-9_1,
© Springer Science+Business Media Dordrecht 2013

Fig. 1.1 Contemporary view of the execution of heretics at Haarlem by the Inquisition (Eytzinger 1585, MPM R25.8)

By now Protestantism had swept the country. Even nobles were beginning to reveal their sympathies. On 5 April 1566, regent Margaret of Parma (1522–1586) was presented with the *Petition of Compromise*, a covenant whereby some two-hundred members of the nobility denounced the Inquisition. It was around this time that the Dutch insurgents came to be mockingly referred to as *Gueux* (Beggars, geuzen), a nickname that has stuck ever since. Margaret was more or less forced to suspend the anti-heresy laws. Protestant, more specifically Calvinist, preachers now openly organized mass gatherings. Beginning on 10 August 1566 in the Flemish town of Steenvoorde, the *Beeldenstorm* (Iconoclastic Revolt) spread across the Dutch-speaking provinces. Central government would not regain control over the rebelling provinces before the Spring of 1567.

Upon hearing of the Revolt, Philip sent Don Fernando Álvarez de Toledo, Duke of Alba (1507–1582 – Fig. 1.2), to the Netherlands at the head of a Spanish Army. Don Fernando was one of Spain's foremost noblemen. Known as the Iron Duke, he was rigidly authoritarian and a fierce opponent of Protestantism. Having arrived in the Low Countries, he ordered the construction of citadels in the main towns and cities, including Antwerp, not so much to keep the enemy out as to keep the towns themselves in check.

He set up the *Council of Troubles* (known as *Bloedraad* or the "Council of Blood" in popular lore), a commission charged with investigating the causes of the revolt and punishing those responsible for it. More than a thousand would be executed as a result of their proceedings. The *Council* condemned inter al. William of Orange and confiscated all of his possessions. William, for his part, responded by putting himself at the head of the Revolt.

Fig. 1.2 don Fernando Álvarez de Toledo, Duke of Alba by Frans Hogenberg (SAA, ICO 12#49)

In 1572, the Sea Beggars succeeded in taking Den Briel and, five days later, Flushing, which meant that they now commanded the Scheldt estuary. Simultaneously, a new rebellion took place, partly in response to Alba's tenth penny sales tax. Alba was able to crush the uprising in the Southern Netherlands, but, for all his effort, failed to subdue the rebellion in the north (Fig. 1.3). The rebellion in the north came at an enormous cost: by 1576 nearly two thirds of all land in Holland had been inundated in defensive actions.

Philip was forced to relieve Alba. He authorized his newly-appointed governor-general, Don Luis Requesens (1528–1576), to seek a negotiated peace. In 1570, a Royal General Pardon was proclaimed and religious persecution more or less ended. This was followed in 1574 by a papal Pardon. With Spain's failure to pay its troops and the death of Requesens in 1576, the Spanish Army of Flanders began to disintegrate as garrisons mutineed. The *Pacification of Ghent* (1576) held the promise of a future without religious strife. The new governor-general, Don Juan of Austria, who realized that he lacked the resources to take other action, agreed to withdraw his troops.

However, in January 1578, after a fresh injection of cash, a resurgent Spanish army launched a campaign to recapture lost territory in the Southern Netherlands. The Spanish reconquest of the Low Countries gained further momentum after Alexander Farnese, Duke of Parma (1545–1592), succeeded Don Juan. In July 1581, the States-General proclaimed the *Plakkaat van Verlatinghe* (Act of Abjuration), renouncing the King of Spain. The situation for the rebels in the south took a turn for the worse as Parma succeeded in taking town after town. Tournai, Maastricht, Bruges and Ghent all opened their gates to the Spanish. Invariably, this prompted an exodus to the north, which slowly drained the south of its life blood.

In July 1584, Parma's troops arrived at the gates of Antwerp. After a siege of barely a year, the city that had become the symbol of the revolt in the south fell to the Spanish.

Fig. 1.3 The Sack of Mechlin by Alba's *tercios* (Eytzinger (1585), MPM R25.8). In the sixteenth century, a town under siege basically had two options: to seek a negotiated settlement or to resist the belligerent army. Mechlin, Zutphen and Naarden sought a settlement, but contrary to Alba's promises, the towns suffered rape, pillaging and murder at the hands of his troops. While the fate of Mechlin frightened Brabant and the Southern Netherlands into submission, the fate of Naarden strengthened the resilience and resistance of Holland and the Northern Netherlands

The victory seemed to have opened the road to the north, but the fledgling Dutch Republic was saved by a political decision by Philip II, who ordered Parma and his troops to Dunkirk to prepare for an invasion of England. Parma's Dutch offensive ground to a halt.

Shortly before his death, Philip II ceded the Low Countries to his daughter Isabella (1566–1633). Together with her husband Albert (1559–1621), the Archduchess would rule the Netherlands as a sovereign in her own right. In practice, the Archdukes perpetuated Spain's policies - with assistance from Spanish troops. One of the first objectives they set was to clean the Scheldt's left bank of States strongholds. This led to a siege of Ostend that lasted a good four years, arguably the longest in modern history. Hence the town came to be known as the new Troy.

In 1607, the war-weary parties concluded the Twelve Years' Truce, during which time an undeniable truth became plain to see for all: the *Leo Belgicus* no longer existed (Fig. 1.4); it had been superseded by two independent nations. In the north, the Dutch Republic would develop into a leading mercantile power. In the south, the Spanish Netherlands began a slow recovery from the pains of war and the resulting exodus. In cities such as Antwerp, international trade had been replaced by a luxury industry. It was hoped, in vain, that the Scheldt would be reopened.

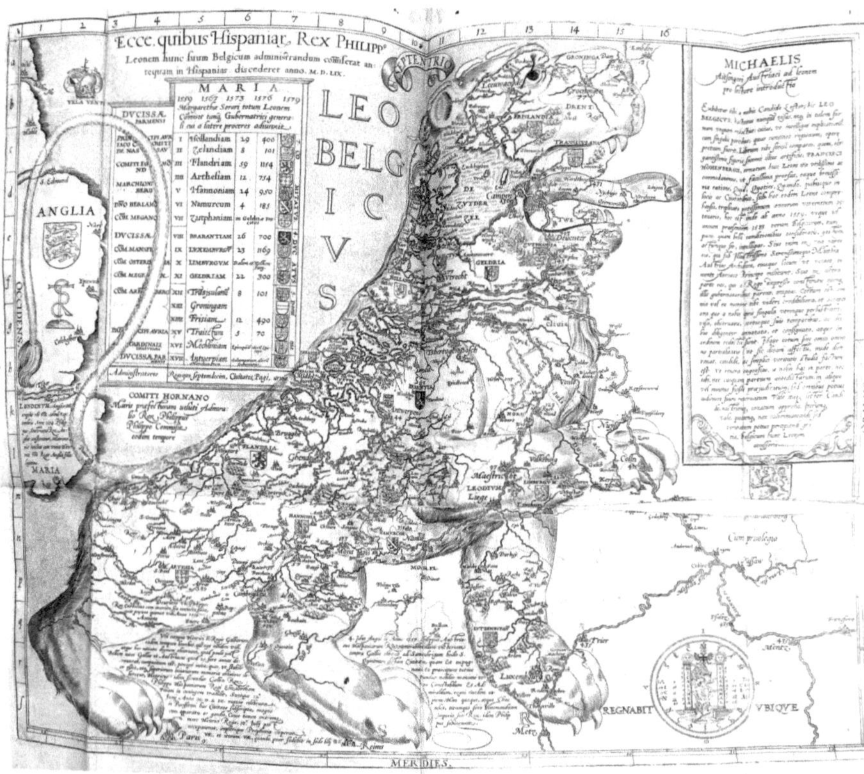

Fig. 1.4 The *Leo Belgicus* at the time of Philip II (Frans Hogenberg in Eytzinger 1583, EHC K 55537)

The Twelve Years' Truce was not renewed (1619), possibly because of the damage the Dutch were inflicting on the trade fleet to the Indies and the Americas. The Dutch-Spanish war now merged into the German Thirty Years' War. Peace would not return to the Low Countries until the signing of the Peace of Munster (1648), which effectively ended the Low Countries' own Eighty Years' War.

1.2 Antwerp in the Sixteenth Century[2]

By the middle of the sixteenth century, the port of Antwerp was thriving and prosperous (Fig. 1.5). Bruges had fallen from prominence more than half a century earlier, while Amsterdam's heyday still lay half a century ahead.

The city government, or magistrate, consisted of two Burgomasters, eighteen aldermen, two treasurers and a *rentmeester* (steward). The aldermen elected the

[2]See inter al. Marnef (1996a,b), Schoups and Wiggers (1998), and van der Stock (1993).

Fig. 1.5 Hans Bol, *Panoramic view of Antwerp and its port*, 1583 (KBC Bank NV, Rockoxhuis)

two Burgomasters: the *binnenburgemeester*, who dealt with judicial affairs, and the *buitenburgemeester* who managed political and military affairs. For important matters or for matters exceeding the local level, a *Brede Raad* (Broad Council) would be convened by the magistrate. The city had been divided into thirteen *wijken* or districts, each of which had two appointed *wijkmeesters* (wardens). The *wijkmeesters*[3] were an invaluable link between the council and the citizenry, translating city policy into rules of daily life. The *wijkmeesters* also commanded the civil guard of their district, assisted by centurions and decurions. They kept lists of the citizens residing in their district, pursued payment on fifth and tenth penny taxes, billeted soldiers, carried out censuses . . .

The sovereign was represented in Antwerp by a sheriff, a deputy sheriff and an *amman*. It was the sheriff's duty to proclaim and enforce laws. He also brought criminals to justice, collected evidence against them and prosecuted them before the *Vierschaar* (High Court). The *amman* was, in the context of a commercial metropolis, the most important representative of the sovereign: he dealt with all civil cases that were brought to court.

As for the Church, Antwerp had been part of the bishopric of Cambrai up to the reign of Alva, when it was made a bishopric in its own right. In the sixteenth century, and despite its booming population, Antwerp was comprised of just five parishes, which made it hard to ascertain whether individuals were regular churchgoers or not.

More than a thousand ships docked at the port of Antwerp annually. The majority were involved in North Sea trade, sailing on England, Holland, Germany and the Baltic. A sizeable number were Portuguese and Spanish vessels, carrying colonial goods to northern Europe.

Prosperity also came from the *jaarmarkten* or annual commercial fairs. At these summertime fairs, Iberians and merchants from southern Germany, who controlled overland trade with Central Europe, would conduct business with one another.

Undoubtedly the jewel in this mercantile town's crown was the *Bourse*, where goods and stocks were exchanged, loans were negotiated and capital was raised for large ventures and expeditions. Among the borrowers, we find Philip II, Henry II of France as well as Mary Tudor and Elizabeth I; among the lenders were the Fuggers, the Welsers and the Antwerp family Schetz.

[3]See Boumans (1965, pp. 52–53).

Fig. 1.6 The Spanish Fury (Eytzinger 1585, MPM R25.8)

In the first half of the sixteenth century, the Antwerp city government showed itself lenient towards Lutheranism, especially if the individuals involved were foreign merchants. Consequently, heretics in Antwerp largely escaped Alba's reign of terror that ruled in other parts of the country. Alba had very pragmatic reasons to accept this: he needed the money Antwerp could lend him to sustain his army. To keep the city in check, he ordered the construction of a citadel on its southern periphery, a project that was finalized within a couple of years.

No event in Antwerp's history was so violent as the Spanish Fury of November 1576 (Fig. 1.6). From the newly constructed citadel, Spanish mutineers went on a three-day rampage through the city, raping, pillaging and killing at will.

After the Fury and the *Pacification of Ghent*, the Spanish troops left the city. Orange proclaimed a *Religious Peace* whereby Calvinists and Lutherans were allowed to worship in designated Catholic churches. The Calvinists' influence would grow until they eventually gained control of the city council.

After a period of relative peace, it became obvious that Farnese was directing his offensive against Antwerp. By the Summer of 1584, his troops were at the city gates. *Buitenburgemeester* Philips of Marnix's hesitation to have the dikes breached and Farnese's decision to construct a pontoon bridge across the Scheldt meant that the city was cut off from the Beggars' supply ships. Its fate had effectively been sealed. On 17 August 1585, the treaty of reconciliation was signed. Upon entering the city, Parma's troops behaved impeccably. Moreover, the terms he had imposed were not onerous. Non-Catholics were given a choice between converting within four years and leaving the city with all their belongings. Despite their magnanimity, these conditions, combined with the bitter winter of 1586–87 and the resulting famine, prompted a massive emigration. Antwerp had gone from riches to rags.

After 1585, the more or less periodic shifting of the frontline northwards and southwards drove the economies of Flanders and Brabant into deep recession. But from the 1590s onwards, Antwerp began its slow recovery. The textile, lace and wood industries, as well as many of the visual arts, flourished, not in the least thanks to the massive expenditures induced by the Counter-Reformation. Quality now took precedence over quantity. Goods were exported to Spain and its overseas territories, and Antwerp merchants also frequently acted as go-betweens for Dutch traders, who in many cases were their emigrated next of kin.[4]

[4]See for example Lesger (2001) on the port of Amsterdam in the second half of the sixteenth century.

Chapter 2
The Coignet Family

When studying mathematical practice in sixteenth-century Antwerp, one figure towers above all others: Michiel Coignet. His career spans virtually all aspects of applied mathematics, which makes him an obvious choice as the central figure in the present study. He was descended from a family of goldsmiths that branched out into the artistic and medical fields. In this chapter, the spotlight is on the Coignet clan.[1]

2.1 The Ancestors

The name Coignet, also spelled Cognet, Quignet or Quinet, first appears in Antwerp in late 15th-century documents. The family most probably hailed from the Lille or Arras area.[2] Etymologically, the family name derives from the French word *coin*, denoting the small chisel used by silversmiths and goldsmiths. This is confirmed by the different variants of the family crest, which always features three downward pointing wedges, or stylized chisels. The coat of arms of the Antwerp branch sported a red twill with three azure wedges (Fig. 2.1).[3] The jeweller Jacob Cunget Peterszone was a progenitor of Michiel's family branch.[4] Around 1511, he married Margriet van der Biest.[5] The couple had at least two sons, Gillis and Jacob II, and three daughters, Margriet, Katlijne and Josijne.[6] Jacob II and Gillis followed in their

[1]On the Coignet family (with an extensive family tree) Meskens (1998a).

[2]Kinget (1989) and van Valkenburg (1988, pp. 72–73).

[3]van Valkenburg (1988, pp. 74–75) and anon. (1863, 2, p. 197).

[4]See Meskens (1998a, pp. 19–21), on the other branches of the family.

[5]SAA SR208, f193*r*-*v*; SR139, f259*v*; van Hemeldonck (1998).

[6]Prims (1948a, p. 103); Schlugleit (1936, p. 51); SAA Pk3473; V144, f266*v*; A4587, f3*v*; SR208, f193*r*-*v*.

A. Meskens, *Practical Mathematics in a Commercial Metropolis: Mathematical Life in Late 16th Century Antwerp*, Archimedes 31, DOI 10.1007/978-94-007-5721-9_2,
© Springer Science+Business Media Dordrecht 2013

Fig. 2.1 Coat of arms of Michiel Coignet (anon. 1863, 2, p. 197)

father's footsteps and became goldsmiths. Jacob was quite wealthy: in addition to several *erfrenten* or rent charges,[7] he also owned several houses.[8] The names of these houses, all of which include the word chisel, refer to the family name and

[7]In sixteenth century Antwerp a rent charge (*erfrente*) was a kind of perpetual lease of a property (in most cases a house) in exchange for which payments were made at fixed intervals and in perpetuity. The seller of a rent charge sold a property, including all rights, benefits and duties attached thereto, to another person, the buyer. On the demise of the seller, his children automatically inherited the rent. When a house was sold outright, the buyer had to pay the entire sum within a certain period of time. For this reason, the buying and selling of rent charges was encouraged among the less affluent. The buyer could at all times repay the whole rent, while the seller could never force the buyer to pay the whole amount. The practice can be seen as the sale of a house (or other property) in exchange for a perpetual rent, thus avoiding usury. Long-term loans in the form of a rent charge did not fall under an episcopal prohibition, because the lender could never be forced into a full repayment. Two papal bulls (1425 and 1455) allowed the sale of rent charges under certain conditions (which were met in Antwerp).

The price-to-rent ratio was expressed in pennies. In Antwerp, it usually amounted to "penny 16", which meant that multiplication of the rent charge by 16 yielded the price of the house. In modern terms, one would say that the interest rate amounted to 6.25%.

The *rentewaarde* or rent charge value was the basis for taxation. It was theoretically equal to the value of the rent charge which could be sold on the house. Since this last value was 1/16th of the value of the house, its theoretical value can be calculated. The rent charge value thus becomes a basis for comparison for real estate prices.

For a complete overview, see Soly (1974, pp. 523–527, 1977, pp. 54–59).

[8]*De Gulden Beytel* (the Golden Chisel), purchased by Jacob in 1526 (van Valkenburg 1988, p. 73), *De Rooden Beytel* (the Red Chisel) in Huidevetterstraat and *De Gheelen Beijtel* (the Yellow Chisel), *De Groenen Beijtel* (the Green Chisel), *De Witten Beijtel* (the White Chisel), *De Swerten Beijtel* (the Black Chisel) and *De Blauwen Beijtel* (the Blue Chisel), all in adjacent Groendalstraat. van Valkenburg (1988, p. 74); van Hemeldonck (1998); SAA SR208, f193*r*-*v*; SR169, f368; SR171, f208.

hence to Jacob's profession. Jacob also owned premises on the Meir,[9] Antwerp's commercial artery, and a summer house in Predikherenstraat.[10] His name also crops up several times in a context of real estate transactions, in and around Antwerp.[11] Jacob died between 2 September 1528 and 8 January 1528 o.s. (=1529).[12]

2.2 Gillis Coignet the Elder

Jacob's son Gillis, also known as Egidius, was born before 1526.[13] The fact that Jacob II and Josijne were acting as guardians of Jacob's other children in 1534 suggests that Gillis was still a minor at the time. This in turn indicates that he was born no earlier than 1514.[14] Gillis was a goldsmith and a maker of astronomical and mathematical instruments.[15] Apart from some details relating to real estate transactions,[16] little is known about his life. Gillis was married to Brigitta Anthonis Hendriksdr.[17]

In 1543, Gillis became a Master with the Guild of St Luke.[18] He died in 1562/1563,[19] leaving behind at least three sons, Jacob III, Gillis I[20] and Michiel, and a daughter, Brigitta.[21] Jacob III became a physician,[22] while Gillis I became a painter.[23] Michiel, for his part, would follow in his father's footsteps, or at least to some extent.

---

[9] van Hemeldonck (1998); SAA SR169, f310*v*.

[10] van Hemeldonck (1998); SAA SR164, f157.

[11] Meskens (1998a, pp. 19–20).

[12] On 22 September 1528, he reached an agreement with his neighbour Peter van Breen on a shared wall; on 8 January 1528 o.s. (=1529), his widow sold a rent on a house in Groendalstraat (van Hemeldonck (1998); SAA SR173, f106*v* and f218).

[13] In 1544, Gillis claimed that he had inherited the house *De Gulden Beijtel* (the Golden Chisel) in 1526. van Hemeldonck (1987, no. 374), SAA SR214, f266 en 269*v*.

[14] van Hemeldonck (1998). Act dd. 26 February 1534 o.s. (=1535).

[15] De Groote (1968e, col.184) and Prims (1948a, pp. 103–104).

[16] van Hemeldonck (1987, no. 374) and Meskens (1998a, pp. 22–30).

[17] De Groote (1968e, col.184).

[18] Rombouts and van Lerius (1874, I, p. 145).

[19] SAA GA4587, f3*v*.

[20] We call him Gillis I (and not Gillis II), because that is how he is known in art literature.

[21] For a detailed family tree, see Meskens (1998a, p. 181).

[22] SAA Pk 3473; Genard (1886, p. 86), Rombouts and van Lerius (1874, I, p. 282).

[23] For a biography of Gillis I, see Meskens (1996c, 1998a, pp. 31–50).

Fig. 2.2 Gillis Coignet, compendium (private collection)

Jacob II seems to have had a good reputation as a physician.[24] In 1582, he became a member of the Guild of St Luke, but it is not known in what capacity.[25] He died in 1614.[26]

In 1561, Gillis I became a Master of the Guild of St Luke (Fig. 2.4). In the second half of the 1560s, he travelled through Italy, visiting, among other places, Florence, Rome, Naples and Sicily.[27] After 1570, he returned to Antwerp, where he would soon employ a number of apprentices.[28] In 1584, Gillis I was appointed Dean of the Guild of St Luke, a position he would continue to hold during and shortly after the Spanish siege.[29] Subsequently, Gillis I moved to Amsterdam, becoming a burgher of the city in 1589. Here he enjoyed considerable success as a painter and would come to influence the art scene considerably. Around 1593–1594, Gillis I moved to Hamburg, where he died in 1599.

---

[24]Genard (1886, p. 86).

[25]Rombouts and van Lerius (1874, I, pp. 282, 339).

[26]SAA Cert 66, Pk 2942.

[27]Dacos (1995, p. 157).

[28]According to the guild books (Rombouts and van Lerius, 1874, I), these were Simon Utens (p. 243), Jaspar van Doorne (p. 256) and Robert Keuls (p. 289). According to Van Mander, the Haarlem painter Cornelis Cornelisz. also worked in Gillis's workshop, presumably towards the end of the 1570s.

[29]Rombouts and van Lerius (1874, I, pp. 284, 289, 295, 300).

Fig. 2.3 Gillis I Coignet's entry in Joannes Vivianus' *Album Amicorum* (Vivianus (1571–), f32*r,* Koninklijke Bibliotheek Den Haag 74F19). The text is a pun on his name: *Heureux Quiniet* (heureux qui n'y est, happy is he who is not there). Joannes Vivianus or Jean Vivien (1543/1546–1598) was a wealthy merchant and humanist, who in 1571 came to Antwerp from Valenciennes. In 1585 he left Antwerp for Aachen (Bostoen (ed.), 1999, p. 27)

2.3 The Loyal Polymath: Michiel Coignet

Fig. 2.4 Valentin Mennher (*top:* Mennher (1563), MPM R50.29; *bottom:* SAA GA4528)

Michiel Coignet was born in 1549.[30] His father Gillis died in 1562/1563 when he was just 13 years of age.[31] Michiel must have received a good education in mathematics, but it is not clear where and from whom. He probably first attended school at age five or six, as was customary in those days. Here, he would have been taught to read, write and count, and would have acquired some French. Michiel's brother Gillis I was apprenticed to the painter Lambrecht Wenslijns.[32] On this basis we may surmise that Michiel, too, was apprenticed, possibly to a schoolmaster, an instrument maker or a goldsmith.

Be that as it may, Michiel was clearly further educated in mathematics. It has been suggested that Valentin Mennher was his tutor,[33] but there is no proof to substantiate this claim. Moreover, there is no reason to exclude some kind of advanced education. Mennher was most probably the only schoolmaster in Antwerp who had the ability to teach higher algebra and the mathematical techniques of trigonometry, two fields in which Michiel would later excel. He may have been the one who initiated Coignet in German mathematical culture.[34] Mennher was born ca. 1521 in Kempten in Southern Germany.[35] On 16 August 1549, he became citizen of

[30]anon. (1863, 2, p. 197). His epitaph, now lost, stated that he died in 1623 on 24 December at the age of 74, which would imply that he was born between 24 December 1548 and 24 December 1549. However, a note in SAA Pk2933, f539, dating from 1568 and concerning his entry into the Guild of St Ambrose, states that he was twenty-two years old at the time, which would put his birth in 1546.

[31]SAA GA4587, f3*v*.

[32]SAA Pk3573; SR386, f93*v*, Rombouts and van Lerius (1874, I, p. 184).

[33]This is asserted by Prims (1948a, p. 104). On Mennher, see Haller (2011).

[34]Coignet's extensive knowledge of foreign literature may also stem from his acquaintance with printers. For one thing, it is not entirely clear whether or not there are family connections between the printer Hendrik Hendriksen, alias Hendrikszoon (son of Hendrik), and Coignet's mother Brigitta Anthonis Hendriksdochter (daughter of Hendrik).

[35]Mennher (1565, f Lvii *v*).

Antwerp,[36] the city he had moved to as an accountant for the Fugger family. In 1565–1566, he became Dean of the St Ambrose Guild of Schoolmasters. He died on 9 August 1570.[37]

Michiel Coignet was admitted to the Guild of St Ambrose in 1568,[38] when he was still residing with his mother in Achterstraat (now Noordstraat). Not long after he joined the Guild, a number of members were expelled because they had sympathized with the Protestant (or, to be more precise, Lutheran) cause. Coignet taught French and what is referred to in the sources as *mathematicam*, most probably "higher" mathematics (see Chaps. 4 and 5).

Around 1570, Michiel married Maria vanden Eynde, the daughter of a bell maker. They seem to have lived in Klarenstraat in the house known as *In de Sonne*.[39] By 1573, they had moved to Vleminckstraat, across *De Gulden Ram*.[40] In fact, moving house was a recurring theme in Michiel's life.

As a young father, Michiel wanted to provide as best he could for his family and therefore combined several professional activities. In 1572/1573, he was appointed as an official wine gauger, a semi-public servant whose job it was to measure the content of incoming wine barrels with a view to establishing how much excise was due (see Chap. 6). As he was paid a modest sum per measurement, the income he derived from this position must have varied from year to year, but it would certainly have constituted a neat supplement to his earnings.

The earliest instrument he is known to have built was an astrolabe (1572),[41] dating from the same period. It was the first in a series of very fine instruments (see Chap. 7).

Business must have been good, because he employed an assistant teacher from 1576 onwards. On 5 May 1576, this assistant was fined for teaching without a licence issued by the Guild.[42] On 14 March 1580, Jan de Rademaker was admitted to the Guild as an assistant to Michiel Coignet, with whom he shared accommodation in *De Gulden Ram* in Vleminckstraat.[43] He was licensed to teach French, arithmetic and accounting. Jan married Maiken Hellemans on 28 May 1587, with Michiel Coignet acting as his best man.[44] De Rademaker gave up teaching between 1584 and 1586 for unknown reasons. He served four terms as Dean of

---

[36]De Groote (1960, pp. 162–164); SAA V146, f108*r*, where it is stated that he came from Augsburg.

[37]De Groote (1967a, p. 28) and Prims (1948a, pp. 104–106).

[38]SAA GA4528, f186*v*.

[39]De Groote (1968e, col. 185) and van Cleemput (1973–1975, p. 113).

[40]SAA A 4550, p. 72; De Groote (1968e, col. 185).

[41]Now in the Kunstgewerbe Museum, Berlin.

[42]SAA GA4829, f38*r*.

[43]SAA GA4550, p. 81; De Groote (1967a, p. 295); De Groote (1968e), De Groote (1970c) and Meskens (1994b, pp. 41–42; 52).

[44]SAA N1480, f107*r*.

the Guild of St Ambrose, in 1596–1598, 1603–1605, 1609–1611 and 1616–1618.[45] De Rademaker is the author of a very successful arithmetic book, first published in 1589.[46]

Also around this time, Michiel got into publishing. His edition of Mennher's *Livre d'Arithmétique* (first published in 1561) appeared in 1573. It is a stereotypical sixteenth-century arithmetic book, beginning with the pronunciation of the number words and continuing up to the extraction of cubic roots. It contains numerous problems, without their solution methods, but with numerical solutions. *Cent Questions Ingénieuses,* which contains the solutions of a hundred questions posed by Mennher in his 1561 book, is very often found in a single volume with *Livre d'Arithmétique,* suggesting that they were commonly sold together.

Some years later, in 1580, he edited Willem Raets's *Arithmetica* (first published in 1567),[47] which contained an appendix on wine gauging. Raets seems to have been a wine gauger from Maastricht, who applied for a similar position in Antwerp (see Chap. 6). Again, this book covers all the problems one would expect to encounter in a sixteenth century arithmetic book, with solutions usually involving the Rule of Three.

The book to which Michiel owes his fame, *Nieuwe Onderwijsinghe op de principaelste puncten der Zeeuaert* (*New Instruction on the most important issues of navigation*), was written in 1579. It was published by Hendrik Hendriksen in 1580 as an appendix to the Dutch translation of Pedro de Medina's *Arte de Navegar*. Both books would be republished in Amsterdam on three occasions by Cornelis Claesz. (1589, 1592, 1598). The French translation of this appendix, *Instruction nouvelle*, was published the next year as a separate book (see Chap. 8).

Michiel became a member of the Guild of St Luke, as the son of a Master of the Guild, in 1581. In the account books, he is described as a *cruyenier en wijnroeyer* (shop keeper and wine gauger).[48] Michiel and his wife Maeyken joined the *Armenbus* of the Guild of St Luke in 1583, most probably at the behest of his brother, the then president of the association (Fig. 2.5).[49]

The previous year, in 1582, he had become a member of the Guild of Meersse (i.e. the Guild of shopkeepers and porters), but without specification of his trade.[50] Since that year, he had lived in a premises called *Het Eeckhorentje* (the little squirrel) on the corner of Braderijstraat and Grote Markt, the town's central market square (Fig. 2.6).[51] The house was taxed at a rent charge value (*rentewaarde*) of 100 gl.

[45]See SAA GA4529 and 4530; De Groote (1967a, p. 295).

[46]See DeGroote (1967a, p. 295, 1968d, pp. 12, 28, 1968e, 1970c) and Smeur (1960a, p. 56).

[47]Although we know from Coignet's comments that the book was published in 1567, there is no known extant copy.

[48]Rombouts and van Lerius (1874, I, pp. 279, 305, 336 and 599). In 1623 Michiel is still mentioned as such upon payment of his *doodschuld* (litt. death debt).

[49]Artesis Dept B Bib 243(4), f29*r*.

[50]SAA FA24, f43.

[51]For a full history of this house, see Asaert (2005, pp. 282–285).

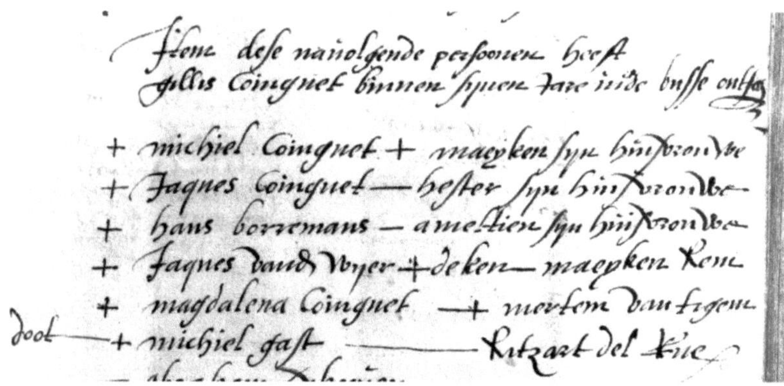

Fig. 2.5 *Bussenboek*, the account ledger of the *Armenbus* of the Guild of St Luke. The *Armenbus* or Poverty Relief Fund may be regarded as one of the first attempts at organized solidarity. In 1583, during his presidency of the *Armenbus*, Gillis I Coignet seems to have persuaded a number of his relatives to join the organization, including his brothers Jacob and Michiel and their respective wives (Artesis University College Antwerp, Dept B Bib 243(4))

Coignet and his wife took over the annual lease of 155 gl. from Anna van Aelst, wife of Tristram Verhoeven, who, in 1581, had rented the house for a period of 6 years from the landlord Jaspar Verryt.[52] Michiel was taxed 2 gl. 10 st. under the "monthly contribution", so the household may be characterized as relatively well off.[53] 1584 and subsequent years were clearly a period of confusion, as is apparent from the often contradictory personal information in documents from this time. And Michiel Coignet was no exception in this respect.

On 8 October 1585, Coignet was admitted to *kolveniersgilde*, one of the six armed guilds that had been established to defend the city alongside the Spanish troops. Until 1591, Coignet served as an "ordinary member", but in 1592 he became an elder of the Guild.[54] Subsequently, up until 1596, he was exempt from guard duty on account of his position as a wine gauger and civil servant.[55] From 1604 onwards, he was categorized by the Guild as "incapacitated", i.e. he was deemed too old to perform any duties, but remained a member.[56] Despite his membership, he was also registered in the civil service registry as "reformed" (i.e. Lutheran).[57] At that

---

[52]SAA Pk2257, f280*r*; SR371, f186*v*.

[53]SAA Fiches VAN ROEY; R2439, f38*v*. The monthly contribution was a special war tax to finance the defences against the Spanish. The amount due was calculated on the basis of households' combined assets. Initially the poor, who made up nearly three-quarters of the population, were exempt. In all there were 12 tax brackets, the lowest of which implied contributions of 1 gl. 10 st, 2 gl. 10 st, 4 gl. and 7 gl. 10 st.

[54]SAA GA4664, f105*r*.

[55]SAA GA4664, f105*r*.

[56]SAA GA4664, unnumbered page.

[57]SAA GA4830(1), f52*v*.

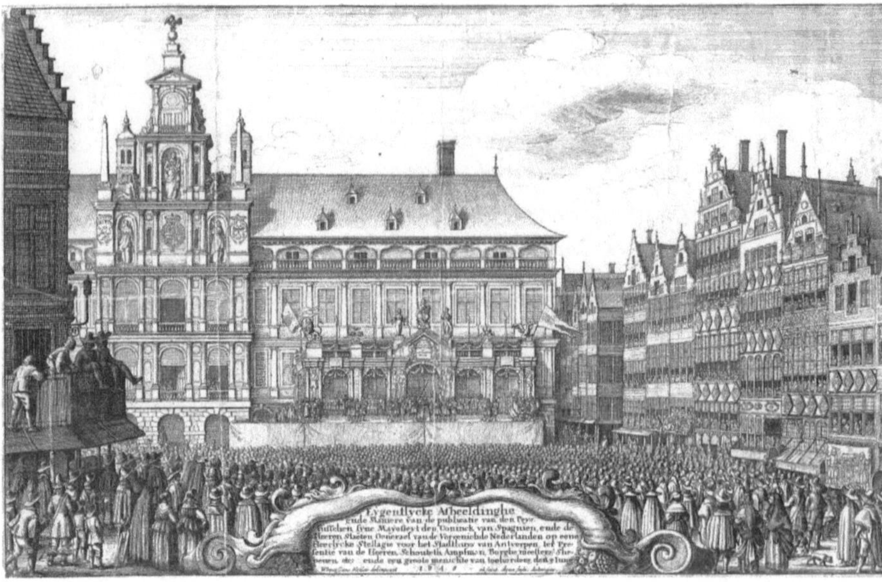

Fig. 2.6 Antwerp's *Grote Markt* (central market square). Coignet's home, known as *Het Eeckhoorentje*, is the corner house at the tip the flag to the right of the town hall (Wenzel Hollar, *Announcement of the Peace of Munster before the Town Hall of Antwerp, 5 June 1648*, MPM-SPk PK.OP.13939, III/H.172)

time, *kolveniersgilde* had already been "religiously cleansed". The few remaining Lutheran and Calvinist members had all vowed to return to the "true Catholic faith".[58]

Between 1585 and 1587, Michiel Coignet may have been the sole wine gauger in the employ of the city, as his colleague Melchior Blommaert resigned his position in order to cultivate an orchard.[59]

In 1586, Coignet bought a rent charge for 60 gl.: a premises in Predikherenstraat consisting in three houses known as *De Bloeyende Wijngaert*, making him in effect the owner of the houses.[60] It is not clear whether he ever actually lived there. In fact, very little is known about his life during this period.

On 13 March 1596, Coignet asked the city council to be relieved of his gauging duties, because he had entered into the service of the Hapsburg court, as a mathematician and engineer to Archduke Albert (Fig. 2.7). Although it is not altogether clear what these positions entailed, Coignet is known to have supervised several projects around Antwerp and to have attended at sieges. His advice was sought during the sieges of Hulst and Ostend[61] (see Chap. 11).

[58]Boumans (1952a, p. 746) and van Roey (1985a, p. 203).

[59]SAA R27; R28; R29; Pk1409, f41*r*-*v*.

[60]SAA SR385, 336*v*–337*r*.

[61]Pinchart (1860, p. 294); ARA Raad van State en Audiëntie 2654, dd. 31 July 1624

Fig. 2.7 Portraits of the Archdukes (François Harrewijn (engraver), Peter Paul Rubens (designer), MPM-SPk PK.OP.18680, IV/H.19 (*left*) PK.OP.18679, IV/H.18 (*right*))

On 13 April 1604, Coignet was awarded 1,000 Flemish pounds (= 6,000 gl.) by the Archdukes in his capacity as cosmographer, but the payment of this grant was postponed until the following year.[62] In 1609, he obtained from the Archdukes a rent charge of 200 gl. on the house of Van Gershoven. The property had been confiscated after the owner, Nicolaas van Gershoven, had emigrated to Zeeland. Shortly after the signing of the Twelve Years' Truce (1612), he lost this rent, as ownership was returned to Van Gershoven.[63]

Around the turn of the century, fate dealt Coignet a series of blows. First his son and apprentice goldsmith Michiel died.[64]

Around the same time, three of his other children succumbed to the plague.[65] A letter from John Hay s.j. to Christopher Clavius suggests that these misfortunes forced him out of his home on Grote Markt and made him move to *De Blauwe*

[62]ARA Raad van State en Audiëntie 2633, dd. 13 April 1604, ANL B2812, f584*v*–585*r*, also f366*r*–367*r*.

[63]Pinchart (1860, pp. 293–294). SAA Proc. Suppl. 1738.

[64]Rombouts and van Lerius (1874, p. 279). Parish accounts Onze-Lievevrouwekerk 1599–1600. de Burbure (s.d.) (SAA Pk2932, f172); Schlugleit (1936, p. 51), van Hemeldonck (1987, no. 379); SAA GA4487, f267*v*. Also Pk2932, f236.

[65]Meskens (2002, pp. 449–450). In all likelihood, the three unfortunate children were Adrianus (1585–?), Joannes (1587–?) and Barbara (1581–?). In 1603–1604, they were still young enough to be living at home. See Meskens (1998a, p. 183).

Engel (the Blue Angel), which he rented for 12 Flemish pounds (=72 gl.) from Jacques and Hans Joos.[66] In 1604, more bad news arrived from Bruges, where his other son Frederik had also died.[67] But that was not the end of Coignet's grief. By November 1605, his wife Maria had also been taken from him.[68] A few months later, in February 1606, Michiel married Magdalena Marinus.[69]

Around the turn of the century, Coignet became involved in the publication of atlases (see Chap. 9). From 1601 onwards, he edited numerous editions of Ortelius' *Epitome* and added an introduction on projections to some editions of Ortelius' *Theatrum.*

Also during this period, he tutored a number of private pupils, including Federico Saminiati and Marino Ghetaldi. Saminiati was a descendant from an Italian merchant's family from Lucca.[70] Marino Ghetaldi (1566?-1626), for his part, was a Dalmatian who led a peripatetic life. After his time with Michiel Coignet, he would go on to study with François Viète in Paris.[71] He stayed in Antwerp for a while and seems to have been involved in some trade with England.[72]

In the Summer of 1623, Coignet made a request to the Infante for a pension. He referred to the house with a rent charge value of 200 gl. which he had been given in 1609, but of which he had never been able to take possession, due to the signing of the Twelve Years' Truce. The Infante granted him a one-off payment of 300 pounds (1,800 gl.) and an annual pension of 200 pounds (1,200 gl.), payable per semester.[73] However, the Infante's decision had come too late. By the end of the year, Coignet had fallen ill and he could feel his end drawing nearer. On 14 December 1623, he had his will drawn up by notary Cantelbeeck.[74] His will stipulated that his wife would receive 400 gl., with the rest of his inheritance being

---

[66]SAA N3570, f233*v*, dd. March 1603.

[67]SAA Pk2942; Cert dd. 1604, f19*v*. This may indicate that Frederik had been taken ill or had sustained an injury during the siege of Ostend, as Bruges served as a hospital city for the wounded.

[68]She was still alive in September 1605 (SAA Cert dd. 2 September 1605).

[69]SAA PR186/1; PR195, f1299 Onze-Lieve-Vrouwkerk: 4 February 1606.

[70]Goris (1925, p. 618), mentions a company called Francisco Saminiati et Cie in Antwerp in 1551. Bennedetto Saminiati, an uncle of Federico's, was an engineer and mathematician. For details, see Arrighi (1960). On the Saminiati family in Antwerp, see also Denucé (s.d.), p. 47 and Subacchi (1995, p. 78 ff.).

[71]In a letter to Coignet (1600), Ghetaldi mentions that *De numerosa potestatum* had been published thanks to his insistence towards Viète and his offer to correct the prints. This letter was, for that matter, included in the book itself (f36). Marino Ghetaldi also maintained contact with Thomas Harriot. The two men were involved in attempts to reconstruct the lost works of Apollonius of Perga (Gatti, 2000, p. 68).

[72]Meskens (1993a, p. 27).

[73]ARA Raad van State en Audiëntie 2653, dd. 11 July 1623, ANL B2925 f432*v*-433*r*. Also de Lettenhove (1909, p. 58).

[74]SAA N 3377 dd. 14 December 1623. Coignet's grandson Michiel Boudaen, his son Antonius and his son-in-law Guillaem Flameng were appointed executors with regard to his children. Gillis Coignet (Jacquessone) and Hendrik van Peenen acted as witnesses.

equally divided among his children, his grandson Michiel Boudaen and the children of Maria and Guillaem Flameng. The boys, Coignet's will reads, should learn a trade, while the girls should exercise themselves in needlework to earn a living. Michiel Coignet died on 24 December 1623 aged 74. He was entombed at St James's Church.[75] Of the ten children from his first marriage, only Antonius survived him.[76] The four children from his second marriage were alive but underaged, so that the administration of the inheritance was entrusted to an appointed *weesmeester* (orphan master). The inventory of Coignet's estate mentions many oil paintings, including portraits of himself at a young age (around 15), his widow, the Archdukes and Marquis Spinola. Rather surprisingly, it also mentions an easel.[77] Michiel Coignet's library was inventoried by Antonius and Gillis II Coignet, but unfortunately this document has been lost. His books and tools were sold for 947-14-6 gl. A cautious estimate suggests that Coignet must have owned some 650 books.[78] A number of instruments, including two sectors, an astrolabe, a ring dial, surveyor's sticks and a magnetic compass, were also sold. In addition, the city of Antwerp owed him 102 gl., while the Archdukes owed him an unspecified sum. On the other hand, 18 gl. was owed to Ferdinand Arsenius for the construction of *pantometers* (sectors).

The Chief Bailiff of the Duchy of Brabant paid his widow 80 and 42 gl. for work he had performed on fortifications.[79] She also received 42 gl. from the city treasury for inspections Coignet had conducted and another 24 gl. for the maps she sold to the city council.[80] She complained to the city about an obligation of 100 gl. they had sold to the city wine gauger Gabriel van Bemel.[81] The widow also appealed to the Infante for a continuation of Coignet's pension, but to no avail. On 31 July, she received 200 pounds (1,200 gl.) as a one-off payment. Magdalena Marinus remarried Christiaan Verdonck, with whom she had at least one child, Barbara.[82] She died on 4 July 1663 and was entombed alongside Michiel Coignet.

---

[75] anon. (1863, 2, p. 196) and Rombouts and van Lerius (1874, I, p. 599).

[76] SAA N 3377 dd. 14 December 1623.

[77] SAA N3378; Duverger (1980–1992, 2, pp. 306–308). No fewer than 30 paintings, of unknown dimensions, are mentioned in the inventory of his home.

[78] Meskens (1994b, p. 200). By comparison, his furniture was sold for 1,172 gl. We have used an average of 1 gl. 10 st. per book, a sum based on the index of the books, barring the *Biblia Regia*, owned by Isabella da Vega. This index was drawn up around the time of Coignet's death, which makes it particularly suitable for comparison. Duverger (1980–1992, 2, pp. 400–461).

[79] ARA Rekenkamer 26282, f14*v* (80 gl. dd. 31 August 1624) and 15*v* (42 gl. dd. 21 January 1625).

[80] SAA R69, f231*v* and f378*v*.

[81] SAA Pk721, 59*v*.

[82] SAA N3403, dd. 8 February 1657; WK983, f165 ff.

Fig. 2.8 Art cabinet (Michiel II Coignet and Forchondt workshop, KBC Bank, Rockoxhuis)

2.4 Michiel's Descendants

Michiel's first born son Julianus (or Julius) may have been an apprentice to the goldsmith Melchior Tremschen in 1588.[83] Julianus's son Michiel was baptized in the Cathedral of Our Lady in 1596, with grandfather Michiel serving as his godfather.[84] Neither Julianus nor Michiel jr. survived Michiel sr.

Maria, born in 1585, married Guillaem Flameng, with whom she had at least four children, two of whom were alive at the time of Michiel's passing. Maria herself passed away before 1623. Guillaem worked at the court of the Archdukes as an engineer.[85] All that is known about his activities in that capacity is that he drew up plans of the fortifications of Damme and Groenlo.[86]

Just one of Michiel Coignet's other children made any claim to fame: his son Michiel II became a successful painter and earned a comfortable living. He married Maria Salet,[87] with whom he had at least six children.[88] In recent years, several

---

[83]Schlugleit (1936, p. 18).

[84]SAA PR2, PR11, f100, Onze-Lieve-Vrouwkerk: 16 October 1596.

[85]Pinchart (1860, p. 295) and Proost (1890, p. 24).

[86]Laurent (1986) and Deys (1988).

[87]SAA N2429, f315, dd. 13 November 1640.

[88]SAA N3403, dd. 8 February 1657; SAA WK 1754.

paintings signed by Michiel II Coignet or attributed to him have come up for sale, and a series of four is in the collection of Lamport Hall.[89] The evidence suggests that he ran a workshop producing relatively small and rather stereotypical paintings. He was also a renowned cabinet painter, who was regularly commissioned by cabinet maker Forchondt (Fig. 2.8).[90]

[89]Fabri (1993, p. 108). Also Rubenianum fototheek V17 Cognet.

[90]Fabri (1993, pp. 87–88). For a detailed description and discussion of such cabinets, see Fabri (1991).

Chapter 3
Peter Heyns and the Nymphs of the Laurel Tree

3.1 The Exiled Humanist

Whereas the body of archival material on the Coignet family is quite extensive, little is known about Peter Heyns's parentage.[1] He was born ca. 1537, the son of Jakob Heyns,[2] most probably the same Jakob who was admitted as a teacher to the schoolmaster's Guild in 1557–1558.[3] Peter, for his part, was admitted into that same Guild of St Ambrose on 1 December 1555, although he himself claims to have opened his girls' school, at the age of 18, on 1 August.[4] It was probably located on Steenhouwersvest.

Around 1558, he married Anna Smits Gielisdr., who gave him at least six children: Zacharias, Anna, Jacques, Catharina, Susanna and Magdalena.

Some four years later, he moved his school to Augustijnenstraat, to a premises known as *De Roode Leeuw* (The Red Lion), across the street from St Andrew's Church.[5] He initially rented the building, but later purchased it. The school, *De Lauwerboom* (The Laurel Tree), attracted girls from across the Low Countries. Peter Heyns became Antwerp's most famous "French teacher" (Fig. 3.1).

---

[1]On Heyns see Serrure (1859–1860), Sabbe (s.d.), Meskens (1996b, 1998–1999), Meeus (1988–1989, 2000, 2003, 2009), Dibbets (1994).

[2]De Groote (1967a, p. 266).

[3]SAA GA4528, f114.

[4]MPM 394, f1–2.

[5]SAA GA4550, pp. 69–85 and 97–99, GA4528 f103*r*–104*r*. On the real estate holdings of the arithmetic teachers, see Sect. 4.8.

A. Meskens, *Practical Mathematics in a Commercial Metropolis: Mathematical Life in Late 16th Century Antwerp*, Archimedes 31, DOI 10.1007/978-94-007-5721-9_3,
© Springer Science+Business Media Dordrecht 2013

Fig. 3.1 Peter Heyns (Ortelius and Heyns 1583, EHC C 41116 (*top*), MPM M240 (*bottom*))

From 1561 onwards, Peter Heyns was active as an author for the Rederijkerskamer *de Bloeyende Wijngaert* (Chamber of Rhetoricians "The Flowering Vineyard"). He provided plays for several of rhetoricians's contests, including the 1561 edition in Antwerp and the 1562 contest in Brussels. As a rhetorician, his writing was quite traditional, but as a poet he imitated French verse and even wrote poetry in French.

Followingthe iconoclastic crisis of 1566, the Lutherans temporarily gained the upper hand in Antwerp.[6] But after the revolt was crushed, all guilds were "religiously cleansed". Peter Heyns was neither prosecuted nor even indicted, yet he seems to have deemed it safer to go into exile. He moved to Germany, to the city of Danzig according to himself, though he is known to have lived in Filzengraben in Cologne, alongside a thousand or so other exiles from the Low Countries.[7] To earn a living, he opened a school there. Gerard de Vivre, an exile from Liège, describes the difficulties and hardship experienced by himself and his good friend Peter Heyns.[8] French schoolmasters frequently got into trouble with the Cologne City Council, as they were seen as propagators of Protestantism, but they were popular with parents because the quality of education they offered was much higher than in German schools.

After Alba's general pardon, Heyns returned to Antwerp. He retook his oath as a schoolmaster on 23 June 1571 and reopened his school. Not for long, though, as all schools were soon ordered to close their doors because of a plague epidemic. Heyns's school remained inactive, not for the compulsory 6 weeks, but for a lengthy four months.[9] Heyns occupied himself in the schoolmasters' guild, and in 1574/1575 was first appointed as Dean, together with Jacob Boon. Five years later, in 1579/1580, he served a second term as Dean, this time alongside Arnout Gillis, followed in 1584/1585 by a third and final term together with Jacob van Houthuys.[10]

[6]See Marnef (1996a).

[7]van Roosbroeck (1968, pp. 138, 255, 305).

[8]Meeus (2000, pp. 306–307).

[9]MPM 394, f1–2.

[10]De Groote (1967a, p. 266).

Fig. 3.2 Drawing from Heyns's *Spieghel der werelt,* 1583 (Ortelius and Heyns 1583, EHC C 41116)

Heyns's school was not far from Plantin's printing workshop. Throughout their lives, Heyns and Plantin remained in close contact with one another. In 1571, possibly at the behest of Plantin, Heyns translated Ortelius's *Theatrum Orbis Terrarum* from Latin into Dutch: *Theatre oft Toonneel des Aerdtbodems.*

When Ortelius was appointed Geographer Royal in 1573, Heyns wrote an occasional poem for Ortelius's *Album Amicorum.* In fact, Heyns is one of the few persons who contributed twice. His daughter Catharina and son Zacharias also made a contribution.[11]

After the Spanish Fury (1576), most of his pupils returned home, so Heyns had a lot of time on his hands. Once again at the behest of Plantin, he condensed the *Theatrum* into the world's first pocket atlas, *Spieghel der werelt* (Fig. 3.2). The work was an enormous success and it would appear in over 30 editions in every living Western European language and Latin. The maps for the first edition were engraved by Philip Galle, who was also the publisher (see Chap. 9).

[11]Puraye (1968, pp. 17, 80, 91).

Fig. 3.3 Jacob Matham after a drawing by Hendrick Goltzius, *Beached Whale*, 1598. (Rijksmuseum-Rijksprentenkabinet (Amsterdam) RP-P-1885-A-9446)

In 1579, Heyns joined the civic militia as a lieutenant under captain Henry Quinget (who, as far as we can ascertain, was no direct relative of Michiel).[12] At the beginning of 1580, Heyns became *wijkmeester* (warden) for the seventh district of the new Calvinist City Council.[13] He would also serve as a clerk to the council of *wijkmeesters*. He was not an unimportant figure in the Calvinist hierarchy of Antwerp, as is apparent from the fact that, in 1582, he (alongside Jan de Pape, Willem Martini and Willem van Schooten) was sent on an embassy to William, Prince of Orange, in Amsterdam.[14] During the siege of Antwerp, Heyns was put in charge of the rationing and distribution of grain.[15]

It goes without saying that his involvement with the Calvinist regime made it impossible for Heyns to stay in Antwerp after the Reconciliation (on emigration, see Sect. 4.9). He fled to Frankfurt, leaving his wife and children behind in Antwerp. Later, his wife would join him, but his son Zacharias stayed on as an apprentice in Plantin's printing workshop. In the first half of 1590, he moved his school to Stade near Bremen. Four years later, in 1594, he moved to Haarlem. In February 1598, a whale washed ashore at Berckhey, near Katwijk (Fig. 3.3). Heyns travelled

[12]Ortelius and Heyns (1579, p. +4*r*).

[13]Ortelius and Heyns (1583, p. *3*r*), also MPM M73.

[14]Gielens (1933, p. 255).

[15]Meeus (2000, p. 306).

to the seaside town to witness the scene with his own eyes. However, the cart ride apparently did him no good, because soon after he was taken ill and by 25 February 1598 he had died.[16]

3.2 The Humanist Playwright

Although Peter Heyns was involved in many other activities as well, for most of his life he ran a school, for which he also produced the teaching materials. As early as 1561, he wrote his first book, on arithmetic. It was entitled *Tot profyte van die willen leeren lustich rekenen met penninghen oft penne. (for the benefit of all who wish to learn to count with pennies (jetons) or with the pen)* and today the only known surviving copy is in the British Library. The title would suggest that he taught his pupils to count with jetons as well as to calculate on paper. This is not the case, however, since the book only deals with counting with jetons. On the other hand, the layout of the household accounts in his book is such that Roman numerals are used on the lefthand pages and Hindu-Arab ones on the right. From the valedictory message at the end of the work, it seems that Heyns was not completely satisfied with his booklet, which he must have completed hastily. He promises to write a revised edition that is "as good as the present one is poor"[17] (see Chap. 5, Sect. 5.1). Contrary to what is often thought, it was not part of *Instruction de la lecture francaise et du Fondement de l' Arithmetique*, which was published by Plantin in 1584.[18] This book contained only nine or so pages on arithmetic, which is far less than the 36 pages of *Tot profyte.* If any of the content of *Tot profyte* was reproduced, it cannot but have been limited to the explanation of the basic operations. This Plantin edition was already a revised edition, so that there must have been others, although none are extant. It was again reprinted, this time in Amsterdam, in 1597, with corrections and augmentations by Heyns's son-in-law Kerstiaan Offermans.

Another of Heyns's schoolbooks that would appear to have been in press by 1567 was the French version of his *ABC* book. The Dutch version, *ABC, oft Exemplen om de kinderen bequamelick te leeren schryven*, was published the following year (Fig. 3.4). The ABC book is an exercise book for early readers and writers. It has only woodcut capitals, but no real text. It has been suggested that Plantin had it made for a young Spanish prince. The initials were purposely cut and the title page was printed in red and black, all of which points at some kind of personal involvement on the part of Plantin (Fig. 3.5 right).[19]

---

[16]Meskens (1996b).

[17]Clair (1955, p. 41).

[18]This title is mentioned by C.P. Burger, who found a copy in the Municipal Library of Frankfurt, which was destroyed during the Second World War. Hitherto no other copy has surfaced. See Smeur (1960a, p. 34).

[19]Meeus (2000, p. 310).

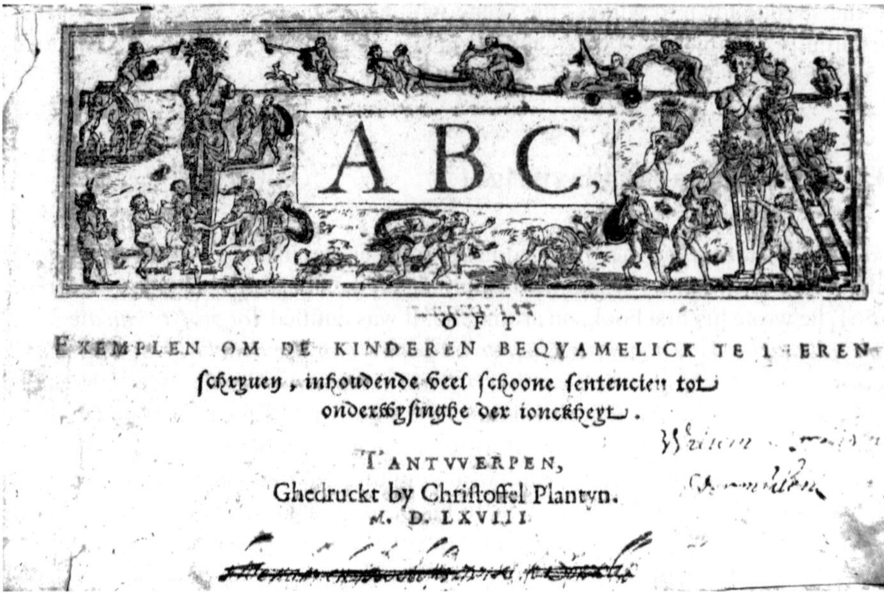

Fig. 3.4 Frontispiece of Heyns's ABC-book (Heyns 1568, MPM R55.24)

Fig. 3.5 Benito Arias Montanus (l.) and Christoffel Plantin (r.) (MPM 4–97)

In 1571, Plantin printed Heyns's *Cort onderwijs van de acht deelen der Fransoischer talen.* Only two quires are known to us. It was republished in 1605 by his son Zacharias at Zwolle. Apparently further editions appeared in 1581 and in 1597.[20] The book is particularly interesting for the study of sixteenth-century French and Dutch grammar and its pedagogy. It influenced quite a few subsequent Dutch writers on French grammar.

As a French poet, Heyns excelled in Arias Montanus's (Fig. 3.5 left)[21] *Divinarum nuptiarum conventa et acta,* for which he wrote fifty sizaines, or six-line stanzas, as captions for the illustrations on Christ's life.

In 1588, Heyns had a French schoolbook published for his German pupils. It bore the title *IIII Dialogues pueriles en Alleman et Francois des quatre saisons de l'an.*[22]

Like the Jesuits a quarter of a century later, Heyns emphasized the didactic purposes of school plays.[23] Plays are relaxing and fun, but they also help improve pupils' fluency, flair and posture. Heyns had previously composed plays for the rhetoricians, so he was acquainted with their structure. But his school drama would appear to have been more modern. Since he ran a school for girls, some of his plays had an exclusively female cast. Heyns believed that women should be cultured, educated, well-informed and learned, but in his plays he also tried to show that it pays to be virtuous or even to merely repent at older age.

Some of Heyns's plays were published by his son Zacharias, including *Le Miroir des Mesnagers* (Amsterdam, 1595), *Miroir des Vevfes; Tragedie sacrée d'Holoferne & Iudith* (Amsterdam, 1596 – Fig. 3.6),[24] *Jokebed. Miroir des vrayes meres. Tragi-Comedie de l'enfance de Moyse* (Amsterdam, 1597). The three plays are all in prose and encompass five acts. They were all performed in Antwerp in the late 1570s or early 1580s.[25]

Heyns announced that his plays would be published "tant en Flameng qu'en Francois" ("both in Flemish [=Dutch] and in French").[26] The Dutch versions of the plays were indeed written and performed, as is apparent from the inclusion of the cast.[27] Listed at the centre are the parts to be played, with the French cast to the left and the Dutch to the right. Whether the Dutch plays were actually printed is unclear.

---

[20]Dibbets (1983, p. 88).

[21]Benedictus Arias Montanus (1527–1598) was a Spanish orientalist and editor of the Antwerp Polyglot Bible. He was ordained around 1559. The Polyglot Bible was published by Plantin (1572, 8 volumes) under the title *Biblia sacra Hebraice, Chaldaice, Graece, & Latine.*

[22]Meeus (2000, p. 310).

[23]On Heyns's school plays, see Sabbe (s.d.) and Meeus (2000).

[24]Dedicated to mademoiselle van Nispen, the widow of Gillis Hooftman (see p. 139), with whom Heyns was on friendly terms. Hooftman's daughters attended his school and performed in Heyns's plays.

[25]The first performance of *Le Miroir des Mesnagers* must have been before 1583. The (exclusively female) cast features names of pupils who left the school in that year (MPM M240, f166).

[26]Meeus (2000, p. 312).

[27]MPM M240, f166.

Fig. 3.6 Judith showing Holofernes' head (Illustration by Peeter vander Borcht in Barrefelt ca.1592, MPM R 24.35)

3.3 The Secretive Humanist

The second half of the sixteenth century saw the rise of numerous religious sects. One of these was known as *Huis der Liefde (Family of Love)* or *Familia Caritatis*,[28] a mystic religious group vaguely comparable with masonic lodges. The *Family* brought together people seeking refuge from strife-ridden religious debates.

The sect was founded by a rich merchant, Hendrik Niclaes (c.1501–1580), known in English as Henry Nicholis, who came from the region between the German town of Münster and the Dutch border. He wanted to create a religious community that drew inspiration from *Theologica Germanica* and *Imitatio Christi*.[29] They felt no need to spread their belief or message, thus avoiding persecution for heresy. Moreover, members of the *Family of Love* outwardly conformed to the predominant religion. With three religions seeking dominance in the Low Countries, it comes as no surprise that the sect's membership in Antwerp included Lutherans, Calvinists as

[28]On this sect, see Hamilton (1981, 1987).

[29]On Hendrik Niclaes, see Hamilton (1981, pp. 24–39). On the sect's sources of inspiration, see pp. 6–23.

well as Catholics. Adherents of the *Family* were said to believe that all things were ruled by nature and not directly by God. They denied the dogma of the Trinity and held that no man should be put to death for his opinions.

In 1564, Peter Heyns was the dedicatee of Marcus Antonius Gillis's *Epictetus,* which was printed by Gillis Coppens van Diest for the publisher Jan van Waesberghe. Marcus Antonius Gillis was the son of the printer Gillis Coppens van Diest (ca. 1496–1572?). He was a member of the intellectual circle around Plantin alongside people such as Johannes Goropius Becanus, Abraham Ortelius and Peter Heyns. He joined the Guild of St Luke as a Master's son in 1572, and became a Master himself just a year later. In 1574 or 1575, he obtained the royal printer's licence. He also worked for Plantin as a translator. In his philosophy, he promotes a humanist rationalist worldview. In the dedication, Gillis tells Heyns that the ethics of the stoic philosophy are easily reconcilable with the values of a Christian society. The core of this philosophy is translated as "liefde der wijsheydt" (love of wisdom), the need to discover the true nature of things. By means of the human intellect, one can attain the wisdom required for leading a good life.[30] Clearly this philosophy more or less ties in with that of the *Family of Love.*

Plantin and Ortelius have been named as possible members of this sect.[31] Although many Antwerp Humanists seem to have maintained ties of some sort with the *Family of Love*, certain prominent intellectuals are notoriously absent from this circle, including Michiel Coignet.[32]

Hendrik Niclaes maintained close contacts with the Antwerp Humanists, but it was Hendrik Barrefelt who was responsible for the daily management of the Antwerp circle. Although they had similar goals, the two gradually grew apart, as Barrefelt was much more tolerant toward the Reformation than Niclaes was. From 1580 onwards, Barrefelt published religious books under the pseudonym Hiël ("The Life of God" – Fig. 3.6).

In the Antwerp circle around *Huis der Liefde*, we find Ortelius and Plantin, and there are some reasons to believe that Heyns was a member or at least a sympathizer. As for Ortelius and Plantin, the reasons for their joining the group – if they joined at all – would appear to have been purely commercial. Looking at the list of (presumed) members of the sect, it is striking how many were acquaintances of Heyns.

Hendrik Niclaes's eldest son, François, lived in Antwerp and was married to Clara Dens. Her brother Adriaan was one of the first followers of Niclaes.[33] Heyns's

---

[30] Vandommele (2011, p. 149).

[31] Hamilton (1987).

[32] Although Coignet maintained relations with both men, the ties would appear not to have been as close as those between Heyns, Ortelius and Plantin. For one thing, Ortelius's *Album Amicorum* contains no entry by Michiel Coignet, but it has two by Heyns and a further two by his children. Coignet is not included in Ortelius's circle of friends by Depuydt (1998). In fact, Coignet's letter to Galileo is the only reference to suggest the two men knew each other (Galileï, 1968, vol. 10, p. 31ff.).

[33] Hamilton (1987, pp. 8–9).

daughter Susanna married their brother Hendrik Dens.[34] Heyns's son Zacharias, despite his father being openly Protestant, was apprenticed to the Catholic Plantin.[35] Hendrik Barrefelt's daughter was a pupil of Heyns's.[36]

Is it unreasonable to assume that the members of this secret society supported each other? Was Heyns a case in point? Did Plantin supply him with work in dire times for teachers? Did Ortelius provide financial support to Heyns so that he could acquire his school premises? Unreasonable or not, the sect having been a secretive society, the answer to such questions will forever remain elusive.

[34]SAA SR396 f156*r* & 160*r*. Meskens (1994b, p. 35).

[35]Meeus (1988–1989).

[36]Sabbe (s.d., p. 23).

Chapter 4
The Arithmetic Teacher and His School

4.1 Introduction

Michiel Coignet began his career as a schoolmaster. Antwerp's schoolmasters, particularly arithmetic teachers, were a mixed bunch, as described in the paragraphs below. The chapter that follows is devoted to arithmetic books, especially those to which Coignet contributed. As hardly anything at all is known about Michiel's private life, perhaps a peek at those of his fellow teachers can shed some light on his existence. After all, he shared with his peers a social status and level of income, as well as certain religious convictions and possibly doubts. As a teacher, Michiel belonged to the social circle he had entered as a young adult, and it was in this environment that he developed into one of Antwerp's and indeed the country's foremost mathematicians. Similar remarks can be made for Peter Heyns.

However, it was also a group that was strongly affected by the religious troubles that swept the country: teachers were compelled to make tough choices that would alter the course of their lives for good. This was a professional category that was torn apart by religious disagreement, much as the country as a whole was.

When we speak of schools, we are inclined to think of more or less official institutions run by a schoolmaster. In the city of Antwerp, such educational establishments were run by the members of the Guild of Saint Ambrose. However, there were also institutions such as Sunday schools,[1] orphanages and Latin schools.

Individual entrepreneurs who opened schools on their own initiative and at their own risk were known as "free" teachers. Especially in mercantile towns such as

[1]In June 1570, the Synod of Mechlin took the decision to establish Sunday schools. These institutions were supposed to cater for the thousands of children who were unable to attend regular schools because they had to work. In Antwerp, bishop Torrentius founded the first such school in 1592, and his commitment to establishing as many schools as possible would be continued by his successors. While the primary focus of Sunday schools was on religious instruction, children were also taught the three Rs at a very basic level. Marinus (1989), Thijs (1987, pp. 110–111) and Put (1990, pp. 104–105).

A. Meskens, *Practical Mathematics in a Commercial Metropolis: Mathematical Life in Late 16th Century Antwerp*, Archimedes 31, DOI 10.1007/978-94-007-5721-9_4,
© Springer Science+Business Media Dordrecht 2013

Antwerp, this type of school flourished. Smaller cities and towns often subsidized a teacher, while in smaller parishes the sacristan would also serve as a teacher.

In its sixteenth-century heyday, Antwerp may have boasted as many as a 150 schools.[2] From the accounts of the Guild, we are able to identify 85 male teachers prior to 1585.[3] This would constitute about 0.5 % of the active male population.[4] The proportion of teachers to residents was about 0.1 %. By the beginning of the seventeenth century, this had dropped to 0.07 %, which was still quite considerable. Although it is difficult to find comparable figures for other towns, it is safe to conclude that Antwerp indeed enjoyed a very high concentration of teachers and schools. In nearby Mechlin and Tournai, the proportion of teachers to residents was around 0.05 %, while in the not unimportant city of Lyons it seems to have been only 0.025 %.[5]

Despite the fact that these schools were private undertakings, the City Council and the Church claimed some authority over them. In addition, each of the five parishes of Antwerp ran their own parish school, also known as Latin schools or *papenschool* (papist school). Their curriculum was not unlike that of an ordinary secondary school, albeit with an almost complete disregard for science. These Latin schools were comprised of six or seven classes, but these were organized entirely differently than today's age-cohort classes: pupils could move on to a higher level at any time of the year, provided that they had passed the relevant exam. Classical authors, the Church fathers and theology made up the mainstay of the curriculum.[6] The scientific knowledge of pupils was often limited to what they had learned at primary school.[7]

---

[2]This number is mentioned in the 1612 edition of L. Guicciardini's *Beschryvinghe van alle de Nederlanden,* where it appears as an addition by Kilianus based on a remark by Carolus Scribani s.j., the rector of the Jesuit College: "150 schools in which youth were taught all sciences and languages" (Guicciardini and Kilianus 1979, p. 123.).

If the estimate takes into account schools where girls were taught needlework and the like, it may well be quite accurate.

[3]SAA GA4528, GA4529, GA4530.

[4]van Roey (1968, pp. 240–242). Van Roey estimates the male population in 1585 at 17,878 (rounded off for our purposes to 18,000) on the assumption that the proportions of male, female and juvenile inhabitants was the same as at the time of the 1698 census.

[5]Lyons had about 60,000 inhabitants in 1571 and just seventeen teachers (Gascon 1971, p. 905; Zeller 1983). Mechlin and Tournai, for their part, had about 25,000 inhabitants (De Vries 1984, p. 272) and 11 teachers each (Briels 1972b, p. 92). A similar proportion is found for the German town of Annaberg (the then second largest city in Saxony), with its population of around 12,000 inhabitants and six teachers in 1540 (Rochhaus 1996, pp. 95–96).

[6]See Nauwelaerts (1978) for a description of books used in the Latin schools of Antwerp and Louvain. By comparison, the Latin school in Annaberger (Saxony) only introduced elementary arithmetic in 1563 (Rochhaus 1996, p. 97).

[7]Beterams (1939), who has published on the *Grootschool* (Latin school) of Mechlin, writes (p. 88): "the worse we picture the condition of the sciences, the closer we approach the truth". Mathematics was taught during a non-compulsory hour, which was, moreover, organized only sporadically, when required for gaining a deeper insight into the writings of the classical authors.

During the period of the Religious Peace (1578–1581) and up to 1585, the parish schools were "protestantized". The schools of the parishes of Our Lady, St Andrew and St James became Calvinist, while those of St George and St Walpurga turned Lutheran.

After 1585, the parish schools were reinstated, but by this time the colleges established by the religious orders were gaining in importance. The Jesuit college had been founded as early as 1575 and would reach its zenith at the beginning of the seventeenth century. In 1605, the Dominicans founded a college, while the Augustinians, at the behest of the City Council, opened their school in 1607. With the exception of the Jesuit college, these institutions played a minor role when it came to mathematics.[8] Towards the end of 1617, after certain difficulties had been overcome, a school of mathematics was opened as part of the Jesuit college. Father Franciscus d'Aguilon s.j. had been destined to become its first professor,[9] assisted by Gregory of Saint Vincent s.j. (1584–1667),[10] a former student of Clavius's.[11] As it turned out, d'Aguilon's untimely death meant that Gregory was appointed as the school's first professor of mathematics. The institution subsequently moved to Louvain in 1621. Laymen as well as Jesuit students could attend classes at the school of mathematics. Among the Jesuits instructed by Gregory were Joannes della Faille (1597–1652)[12] and Theodore Moretus (1602–1667).[13]

4.2 The Guild of St Ambrose

On 19 May 1530, the schoolmasters were granted the privilege to unite in a Guild.[14] Among the most important regulations of the guild was the prohibition for non-citizens to teach either Dutch or French. Teachers were also offered protection against parents defaulting on tuition and against unfair competition.

Following an Imperial Edict of 1546, which was primarily concerned with religious convictions, all schools were to be inspected by a *scholaster*, an ecclesiastical dignitary and chairman of the local Church education council, and by two superintendents, who were civil servants. Anyone intending to establish a school had to obtain their permission. The same edict stipulated that schoolmasters could

---

[8]On the Antwerp Jesuits' mathematics school, and its teachers and students, see Bosmans (1928), Ziggelaar (1983), van Looy (1979, 1980, 1984) and Meskens (1997c, 2005a).

[9]van de Vijver (1980, p. 266) and Ziggelaar (1983).

[10]van Looy (1979, 1980, 1984), Meskens (1997c) and Dhombres and Radelet-de Grave (2008).

[11]Ziggelaar (1983, p. 49).

[12]For a biography of Joannes della Faille, see Meskens (2005a).

[13]Bosmans (1928).

[14]For detailed accounts of the guild, see Bourland (1951) and Poffé (1895); on the schoolmasters, see De Groote (1967a, 1968d) and a number of minor studies by Floris Prims in his collections *Antwerpiensia* (see De Bougnge 1952).

only use books that had been explicitly sanctioned. Remarkably – or perhaps not so remarkably considering their inconspicuous nature – arithmetic books were not mentioned.

The regulations of the Guild were strictly enforced, as is apparent from the regular entries into the accounts of payments of penalties "for certain offences".

These offences need not have been serious. Moving the school without notification, for example, would have resulted in a fine of 1 gl., as Peter Goossens discovered.[15] Similarly in 1587, after Michiel Coignet's school had moved premises, his wife forgot to notify the deans or to hang out a sign, an infringement that cost her 3 gl.[16] Penalties were also imposed for organizing mixed-sex education. Hilde Smeyers was fined for using "inappropriate" books and Hans vanden Bossche for employing an assistant teacher who had not taken his exam.[17]

Punishment for opening a school without licence was harsher. Fines for such infringements could be considerable, and could result in financial ruin. When Valentin Mennher opened his school, he was quickly ordered to shut down, only to re-open after having obtained the Guild's approval. He was ordered to pay a (relatively low) fine of 2 st.[18] In 1566–1567, Pierre Savonne, author of arithmetic books, was forced to close down his school, for that same reason, i.e. not being a Guild member. He apparently left the city after the decision, but returned in 1574 and applied for admission to the guild.[19]

Opening a school without permission was no laughing matter in the eyes of the Guild. On 11 May 1594, Thomas vanden Bossche was brought before the *Vierschaar* and was fined 100 gl. for such an offence.[20] In that same year, he was admitted to the Guild on the condition that he would not teach more than "15 or 16 maidens", a number which may not have sufficed to make a decent living.

From 1560 onwards, the Deans were required to draw up annual lists of Guild members, with a specification of the subjects they taught. After the Iconoclasm of 1566, the Guild regulations became stricter, as new admission rules and a new oath, which called for allegiance to the Catholic Church, were introduced. Henceforth teachers had to renew their licence annually. The Deans were further encouraged to visit all schools around 1 April to ascertain that teachers complied with the regulations. In 1568, a commission was formed to investigate the conduct of teachers "at the time of the Beggars". The commission suspended anyone who had shown sympathy to the Lutheran or Calvinist causes.

[15] SAA GA4529, f87*r*.

[16] SAA GA4529, f115*v*. Remarkably, neither Michiel Coignet nor his wife Maria vanden Eynde are mentioned as guild members in that year. Perhaps they employed an assistant who taught on their behalf.

[17] SAA GA4529, f115*r*–116*r*.

[18] SAA GA4528, f68*v*.

[19] SAA GA4528, f180*v* and GA4529, f30*v*. This may also be an indication that he was a Protestant.

[20] SAA V1104, f104*r*.

In 1577, the Guild was granted a new charter. From then on, no person was permitted to teach languages or to advertise as a schoolmaster without the consent of the Guild. Furthermore, members were required to show their teaching certificate (the so-called "admission letter") at the request of the commissioner.

With the Religious Peace of 1579 came yet another regulation. All active teachers were required to renew their oath on 30 October, which most of them did, including Michiel.[21] Most importantly, applicants no longer needed to gain approval from the *scholaster*; permission from the superintendents henceforth sufficed. Once this permission was obtained, applicants had to pass an exam before a jury of four active teachers. Furthermore, minimum ages were introduced: 25 years for schoolmasters, 20 for schoolmistresses.

The first *sitdagh* (examination day), under the Deans Peter Heyns and Arnout Gillis, was held on 14 November 1579. For the first time in the history of the Guild, prospective members were being called in to take an exam. Eight candidates were invited for an examination in the presence of four elders and "*Mr Michiel Coignet, Cijffermeester ende excellent Mathematicus*" (reckoning or ciphering master and excellent mathematician); a testimony to the reputation that Michiel Coignet had earned by the age of 30. Coignet was included on the jury specifically to test applicants' knowledge of arithmetic. All of the applicants were, for that matter, accepted to the Guild.[22]

After 1585, and the ensuing epuration, life returned to normal for the Guild. Most pre-1585 regulations, not relating to religion, were upheld. Scholaster Reinier Bervoets drew up a list of teachers who were permitted to reopen their schools. Some were expressly forbidden to do so. Moreover, the number of schoolmasters and schoolmistresses was capped at 40 each.

4.3 To Teach and Maintain: The Pupils of the Reckoning Masters

The education of Antwerp merchants' sons started between ages 5 and 7, when they were enrolled at a school, either in their home town or elsewhere, and it would last to between ages 10 and 12.[23] At school, they would learn to read and write, and they also received some religious education and possibly some arithmetic.

After this formal education came the *Wanderjahre* for the sons of the more affluent international merchants: a time of travelling and working abroad. In the course of the seventeenth century, this tradition made way for a more formal type of education at a college.

[21] SAA GA4550. The list is published in Prims (1951a, pp. 214–217).

[22] Bourland (1951, p. 46) and Serrure (1859–1860, pp. 366–377).

[23] For examples (the children of the Della Faille family and their agent), see Brulez (1959, pp. 230–231 and pp. 499–567), Meskens (2005a), also Meskens (1996a, p. 140).

The reputation of Peeter Heyns's school extended far beyond the city gates.[24] In his school, *De Lauwerboom*, which was located in Augustijnenstraat, girls learned how to to read and write Dutch and French, and to count and calculate. Between 1576 and 1585, 461 girls, 247 of whom were boarders, enrolled at his school.[25] Many of the pupils came from cities in Brabant, Flanders, Holland and Zeeland, the most affluent regions of the Seventeen Provinces. They belonged to the upper social strata, many of them daughters of senior civil servants of local and central authorities.

We have very little quantitative information about class or school sizes (which in many cases amounted to the same thing). School sizes in fourteenth-century Florence ranged from 25 to 200 pupils.[26] In the German town of Ulm, ca. 1620, the number of pupils similarly varied between 16 and 200.[27] Valcooch, in his 1591 publication entitled *Regel der Duytsche schoolmeesters* (Rule of the Dutch schoolmasters) mentions schools of up to 400 pupils.[28] Schools in Antwerp seem to have averaged around 50 pupils. Heyns, in his book *Le Miroir du Monde* (1579), wrote that it was his duty "d'enseigner et entretenir à ma table une cinquantaine de jeunes filles" (to teach and maintain at my table some 50 young girls). His accounts suggest that he was referring to boarders only. They also suggest that pupils who entered his school in 1576 on average stayed there for 13.9 months. This may well be too low an estimate, as the *Spanish Fury* caused some pupils to return home temporarily and to pick up their education again some months later.[29] The school apparently struggled to recover from this considerable setback. In 1578, Heyns still lamented that his school was in decline.[30] On the basis of the above figures, one may assume that around 58 pupils attended Heyns's school at any one time. There is however a considerable difference in the length of time pupils spent at the school, with some attending for just 3 months and others staying on for over 3 years. Of the 116 pupils who entered the school in 1576, we know 59 spent at least a year there.

The school run by Jan Borrekens, Peter Heyns's brother-in-law, must have been similar in size. Shortly after his death in 1597, an inventory was drawn up of unpaid tuition and boarding fees. Unpaid fees for 36 pupils ranged from 10 st. to 21 gl.

We have no indication of how many pupils attended Coignet's school, but the fact that he employed an assistant teacher suggests that either it must have been very successful or he did not spend much time teaching himself.

---

[24]On Peter Heyns see Dibbets (1994), Meeus (2000, 2009) and Meskens (1996b).

[25]Sabbe (s.d., p. 18) and MPM M394.

[26]van Egmond (1976, pp. 105–106).

[27]Hawlitschek (2005, pp. 215–216).

[28]De Planque (1926, p. 31). The Latin schools of Delft, Haarlem, Zutphen and Rotterdam had an average number of pupils of around 50; Utrecht had a large school, attended by more than 400 pupils. Bloemendal (2003, p. 20).

[29]MPM M394 f22*v*. One example is Anneke Geeraerts, who "went home on 15 November because of the rioting and returned on 14 May 1577."

[30]Serrure (1859–1860, p. 321).

In 1584, all teachers were required to take a new Protestant oath.[31] All those who complied were inventoried, with specification of the subjects they taught. None of the seventy-one female teachers taught anything resembling arithmetic. By contrast, 51 of the 76 – or two-thirds – of the male teachers claimed to teach arithmetic.[32] It is remarkable that forty-three taught the same combination of subject matter: arithmetic and French.

This combination may seem odd at first sight, but actually it is not given Antwerp's mercantile nature at the time. These were, after all, the subjects that were of greatest use to a merchant. The importance of arithmetic is obvious, and is attested throughout this book. French was equally important, not only to merchants, but also to civil servants.[33] In the County of Flanders, a proportion of the population had always spoken French; in fact, during the Middle Ages, it had been an integral part of France. Since the Burgundian Dukes had become sovereigns of what was to become the Seventeen Provinces, the civil service had used mainly French in its correspondence. This was, moreover, the time when French literature was gaining prestige in the Low Countries and the German Empire. The approach to teaching French was very pragmatic, with an emphasis on practical vocabulary (e.g. in a business context).

Michiel Coignet and Peter Heyns were two of the schoolmasters who taught this combination of subjects. Yet, even a superficial comparison between the two reveals a marked difference. Arithmetic at Peter Heyns's school was limited to the rule of three. On the other hand, Heyns was a gifted playwright, who wrote successfully in both French and Dutch.[34] His plays were performed not only by his pupils, but also at the Antwerp *Landjuweel* (a contest for theatre companies) in 1561. Michiel Coignet – as will become apparent in the next chapter – was well versed in mathematics, to say the least. At a very young age, he was already revising French and Dutch arithmetic books that used advanced mathematics, such as third-degree equations.[35] Nothing is known about his ability as a language teacher apart from the fact that he was fluent in French.

Hence, the accuracy of these lists would appear to be debatable. Perhaps some schoolmasters considered accounting as part of arithmetic. In other words, it is unclear what precisely the content of their arithmetic courses was.

---

[31] SAA GA4550, pp. 245–256.

[32] Fifty-one arithmetic teachers, including as many as 22 after 1585, seems quite a substantial number. Florence, even in its heyday, never had more than seven arithmetic teachers. van Egmond (1976) was able to identify no more than 39 arithmetic teachers in Florence for the period 1300–1500.

[33] See Swiggers and De Clercq (1995).

[34] Meeus (2003).

[35] By way of illustration, there would appear to have been just two sixteenth-century arithmetic teachers in Germany who occupied themselves with such equations: Michael Stifel and Johann Jung (Schneider 2002, p. 11).

In 1585, a similar inventory of teachers and their subjects was drawn up. Here we find just 43 teachers,[36] 22 of whom were also arithmetic teachers. More than half of the arithmetic teachers had given up their profession, possibly because they were about to emigrate. Comparing the two lists, one can see that 19 of the latter 22 were also on the 1584 list. Remarkably, one teacher had it noted that he did *not* teach arithmetic. None of the 65 female teachers taught arithmetic, though that was about to change. During the first half of the seventeenth century, nearly a quarter of the admitted females taught some form of arithmetic.[37] This period seems to have been quite exceptional, though, as in later periods the number of female arithmetic teachers dropped to about 5 %. One possible explanation for the relatively large proportion of females is that there was a shortage of male arithmetic teachers. Of all women admitted between 1585 and 1620, 15 were restricted to teaching counting with jetons, seven were allowed to teach arithmetic, and three could teach ciphering. Soetken Cornelis and Isabelle Gramaye were permitted to teach ciphering up to the rule of three[38] (see Appendix A.2).

4.4 A Schoolmaster's Income

Antwerp provided no public support to schools, with the possible exception of Sunday schools and colleges at the beginning of the seventeenth century. The income of a schoolmaster, therefore, depended heavily on the fees he received. Despite its lack of financial sponsorship, the city council did try to regulate certain aspects. It demanded, for example, that parents would pay up front for either 1 or 3 months regardless of whether their child subsequently dropped out or not. Furthermore, any schoolmaster who accepted a pupil who had not properly paid his or her previous teacher risked a fine.[39]

At least three historical events illustrate how uncertain the income from teaching could be. First, in 1571, a pestilence epidemic caused the authorities to close down schools for several months. Five years later, the *Spanish Fury* drove many pupils out of town. And obviously the events of 1584–1585 also spelled financial disaster for teachers.

Education, it seems, never was the road to riches.[40] This becomes clear from the register of the monthly war tax, known as the *quotisatie* (see p.17). If taxed

[36]Or, more accurately, forty-two, because Peter Simons died in 1585 around the time the list was drawn up.

[37]Put (1990, pp. 193–195).

[38]SAA GA4529, f208*r*, GA4530, f13*v*.

[39]Put (1990, pp. 174–175) and Bourland (1951, p. 57).

[40]In places where schoolmasters were paid by the local council, like the German town of Ulm, the stipend was not very high. In 1520, a teacher was paid 10 Ulm guilders; by 1543 this had risen to 13 gl. For the sake of comparison, 1 Ulm guilder equalled about 1 gl. 4 st. in Antwerp. Hawlitschek (2005, pp. 218–219).

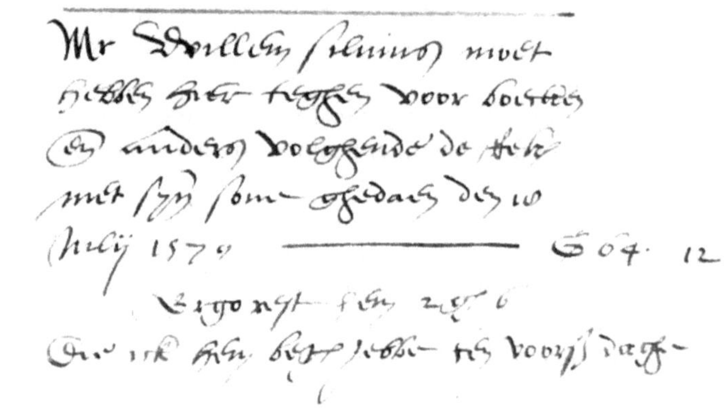

Fig. 4.1 Willem Sylvius's account in Peter Heyns's ledger (MPM M394)

at all, we invariably find the schoolmasters in the three lowest brackets. Two schoolmasters, Melchior van Elselaer and Arnout Hesius, were taxed 4 gl.; four others, including Peter Heyns and Michiel Coignet, were required to pay 2 gl. 10 st.; and 21 others paid less than 2 gl. No fewer than 58 schoolmasters were not taxed at all because their income was too low. In other words, they were categorized as "poor".

Still, it is difficult to assess the income of schoolmasters. The 1530 charter fixed some fees,[41] but inflation would soon render these amounts insufficient, making comparison virtually impossible. Teaching reading and religion would earn a schoolmaster 20 stivers per annum per pupil, while subject matter such as Cato or Aesopus was worth 28 stivers a year, and writing 32 stivers. There were no fixed sums for instruction in arithmetic or geometry: the schoolmasters were free to ask "whatever the good people were willing to give".[42]

Peter Heyns's accounts shed further light on the income of a schoolmaster. The average tuition fee – including arithmetic – amounted to 10 gl. per year for a non-boarding pupil, and to 50 gl. yearly for boarders.[43] Assuming that the fee for non-boarding pupils approximates to the net takings from teaching, then Peter Heyns would have accumulated an income of 400–500 gl. However, he was apparently not always paid in cash, but also in kind. Willem Sylvius, the printer, paid in books for the tuition of his daughters Mynken and Susanna (Fig. 4.1). Other

[41]Bourland (1951, p. 10).

[42]Bourland (1951, p. 52), citing the privilege of 1530 (Capsa Dominorum 14 Nr.25 Scholastria (1400–1700), pp. 6–11). In Gouda, the emigrated Antwerp schoolmaster Jacob vanden Hove charged 8 stivers per month for Dutch only, 12 stivers for Dutch and French, and 20 stivers for a curriculum of Dutch, French and arithmetic (Briels 1972c, p. 295).

[43]Similar amounts were charged by other teachers: 33 gl. per annum in the case of Jacob Huyssens (SAA Cert5, f205) and 11 gl. 9 st. per quarter (= 45 gl. 4 st. per annum) for the services of Andries Roosen (SAA N1478, 17 September 1581).

parents paid in butter, beer, sugar, wine, soap, vinegar, ... In view of the earlier remarks regarding the monthly war tax, an income like Peter Heyns's should be regarded as the upper limit.

During a trial against Karel Strytberger in 1584, the defendant claimed that his apprehension had lost him pupils worth annual fees of 269 gl. in cash and 50 gl. in in-kind payments.[44]

There are even examples of poor arithmetic teachers, to the extent that they had to rely on support from the guild. Jan van Enghelen and Noel Morel, for example, were exempt from paying their annual fee on grounds of their financial circumstances.[45] In 1574, the entire collection made on the night of St Thomas went to poor or needy schoolmasters and -mistresses.[46]

4.5 Other Jobs

Although running a school would appear to have been a full-time occupation, financial uncertainty drove teachers to take on one or more additional jobs. With the possible exception of Michiel Coignet, no schoolmaster in Antwerp seems to have moonlighted as a surveyor. And Michiel Coignet is also the only one who appears to have worked as a wine gauger (see Chap. 6). One may assume that, in a time of high illiteracy, teachers also earned some money writing and/or reading out aloud letters. Carel Bailleul, for example, frequently served as an intermediary between citizens and the city council or the *vierschaar*.[47] Schoolmasters commonly acted as bondsmen before the *vierschaar*, as trustees upon deaths or as executors of last wills. Undoubtedly their literacy will have made them particularly suited to these roles. It was not uncommon, either, for teachers to make calculations on behalf of craftsmen.[48]

Even the Guilds would call on the services of schoolmasters. In 1603, Jaspar de Craeyer "wrote" (possibly the reference is to calligraphy) the statutes of the Guild of St Luke, which represented painters and other artists.[49] Such activities

---

44 SAA Proc. S208 (1585).

45 SAA GA4528, f94*r*.

46 SAA GA4529, f30*v*.

47 SAA V1404 f202*v*, V1407 f142*v*, SR400 f213*v*, SR406 f92*v*-93*r*, SR414 f278*r*, SR424 f40*r*.

48 van Egmond (1976) describes how arithmetic teachers in Florence were paid to calculate the wages of masons (who were remunerated in accordance with the length (or volume) of walls they had built). In Antwerp, this task appears to have been entrusted to the city surveyors (see for example SAA Pk2243).

49 Rombouts and van Lerius (1874, I, p. 320).

could sometimes lead to more permanent appointments. Jacob Huyssens worked as a steward for the Chaplain of St Andrew's Church,[50] while Godevaert Lens took care of the administration of Margriet Suys's possessions.

Fig. 4.2 Martin vanden Dycke (MPM A3290 (*top*), SAA GA 4529 (*bottom*))

Furthermore, it was but a small step from administrative work to the office of notary. In 1582, no fewer than eight schoolmasters also worked as notaries. Martin vanden Dycke (Fig. 4.2) was appointed notary on 10 February 1579 by the Council of Brabant. He first served for a year under notary Anthonis van Male and was subsequently employed by notary Joos Boekhorst, for whom he was still working in 1582. He did not abandon his school though; on the contrary, in 1590–1592, he served as Dean of the Guild of St Ambrose.[51]

In 1572 Michiel Coignet became an official wine gauger, a position he would hold until 1596 (see Chap. 6). As far as we have been able to ascertain, he was the only schoolmaster also to be appointed as a City wine gauger. Other teachers may well have worked as unofficial gaugers.[52] There may, after all, have been other reasons than tax purposes to have the volume of a barrel measured, but very little is known about such unofficial measurements. Michiel Coignet does however write that "the merchants are worried because when one [gauger] has gauged, the other [merchant] will have it [=the barrel] gauged again because one wants to check on the other".[53]

The arithmetic books by Raets-Coignet (1580) and Vanden Dycke (1600) both contain an appendix on gauging. Mennher (1556 and 1565) discusses the construction of the gauge and some of the exercises in these books also deal with gauging. Clearly, then, at least some arithmetic teachers were knowledgeable on the subject of gauging, which suggests they may have acted as unofficial gaugers.

[50] SAA SR 362.

[51] De Groote (1961, pp. 168–170), and SAA GA4529, f156*r* and 164*r*.

[52] In his seminal article, Menso Folkerts suggests that wine gaugers were commonly arithmetic teachers because most of the gauging manuals had been written by them. However, this appears not to have been the case in sixteenth-century Antwerp, at least not insofar as the city's official wine gaugers are concerned. See also Chap. 6, Folkerts (1974) and Meskens (1994a).

[53] SAA Pk1409, f49*r*.

Michiel Coignet had several other sources of income, including a workshop for mathematical instruments (see also Chaps. 6 and 7). So it should come as no surprise that, as of 1576, he needed to employ an assistant teacher.[54] Coignet apparently continued to run a workshop up to his death, while he eventually seems to have restricted his teaching to private tutoring,[55] possibly to explain the workings of his instruments.

Peter Heyns was involved in official administrative roles between 1579 and 1585.[56] During this period, he acted repeatedly as a *wijkmeester* (warden) or clerk on behalf of the board of *wijkmeesters*, for which he was paid 200 gl. per annum (about half his income as a schoolmaster). It is no coincidence that Heyns worked as a *wijkmeester* during this period, as it is a testimony to his protestant sympathies and to the faith the Protestant city council put in him.

Many schoolmasters seem to have been involved in certain kinds of mercantile activity or to have maintained close professional relations with merchants.

Olivier de Cuyper was registered as a merchant in the *poortersboeken*.[57] In 1589, Melchior van Elselaer complained about unpaid bills for a consignment of dried herring.[58] Other schoolmasters had wives who were involved in trade, e.g. Melchior van Aelst's wife, Maria de la Feuer, who ran a grocery shop.[59]

These activities were aimed mostly at the local market. Nevertheless, some arithmetic teachers did not restrict themselves to trade within the city. Aert de Cordes, for example, was a wool merchant. In 1577, he embarked on a business trip to England.[60] He seems to have given up his school after 1585, but remained active as a merchant. In 1586, he imported wool from Lille.[61] As schoolmasters were essentially entrepreneurs, they were free to close shop, be it temporarily or permanently, whenever they saw fit, which is precisely what many of them did.

Hercules de Cordes's main activity was not in education; he was more successful as an accountant in the employ of local and foreign merchants.[62] In 1570, he stayed

[54] SAA 4839, f38*r*. On 5 May 1576, an unnamed assistant to Michiel Coignet paid a fine of 1 gl. for teaching without having obtained permission from the Deans. In 1580, Jan Rademaker (ca.1560–1640), the author of one of the most successful seventeenth-century arithmetic books, served as Coignet's assistant teacher.

[55] Pinchart (1860, p. 210), cites a document from 1601 (ANL, Chambres des Comptes, 279) in which it is asserted that Coignet had begun his private tutoring activities 24 years earlier, i.e. around 1577.

[56] Sabbe (s.d., pp. 12–13).

[57] In 1577, he vowed for a Portuguese merchant who resided in Antwerp (SAA SR350, f113*v*).

[58] SAA Cert50, f410*v*-411*r*, dd. 23 October 1589.

[59] SAA W35 (1582).

[60] SAA Cert39, f56–58.

[61] SAA SR386, f640*r* and 646*r*.

[62] He served as an accountant to, among others, Robert Coels, a market vendor, Adolff Triermann, a merchant from the East, and Herman Boelman, whose wedding in 1560 he attended. SAA Cert 20, f312–315, Cert 31, f399*r*-*v*, Cert 33, f133*v* and Cert 44, 23*v*.

at Lübeck, where he kept the accounts of Herman Boelman.[63] Although his home, a house called *De Penne* in Breestraat, was estimated at just 45 gl. rent value, Hercules de Cordes must have been well to do. He, together with four other investors, held a stake in a company established for a period of 6 years for the purpose of conducting trade with Russia.[64]

It seems that teachers such as Aert de Cordes and Hercules de Cordes were guild members only in name (and by their contributions). One benefit of their membership was that they will have been exempt from serving in the city guard.[65] The Guild, for its part, will have benefited financially from such "sleeping" members, which explains why membership records were kept secret.

4.6 Relationships Between Schoolmasters

In the late sixteenth century, the profession of schoolmaster was not often passed on from father to son. One of the few sons of a schoolmaster also to work as a teacher during these troubled times was Peter Heyns. However, the picture subsequently changed. By the beginning of the seventeenth century, many schoolmasters' sons entered the teaching profession themselves, including Jan Borrekens the Younger, Jeronimus van Blinckvliet the Younger, Jacques Pigaiche, Jan Hesius and Lennart de Raeymaker, not to mention the many sons of those who emigrated north, such as Jacques Heyns and Pieter van Meldert. Sometimes a teacher's widow would continue to run a school after the death of her husband until her son was old enough to take over. Sebastiaan Cuypers, writing to the city council, boasted that no fewer than 13 or 14 of his children or kin had taught at Antwerp schools.[66]

Marriages among teachers' families seem to have been common in Antwerp, so that the ties between them grew ever closer.[67]

No direct relatives of Michiel Coignet are known to have been schoolmasters, nor did any of his daughters marry schoolmasters. His offspring rather pursued careers in engineering and in the fine arts. We have been able to identify a couple of distant cousins, descendants of Christoffel, the progenitor of a different branch of the Coignet family than that to which Michiel Coignet belonged. Matthijs and his son Christoffel Quinget were schoolmasters before 1580.[68]

---

[63]De Cordes complained that Boelman kept his "records" on scraps of paper and that, consequently, he had to extend his stay in Lübeck, which meant he was unable to do the accounts of other clients in Antwerp. On Boelman and his associate Gerard Gramaye, see Wijnroks (2003, pp. 73–80).

[64]Wijnroks (2003, p. 71).

[65]Put (1990, p. 95).

[66]De Groote (1967a, p. 247) and De Groote (1968d, p. 7).

[67]See Meskens (1996a).

[68]SAA V1393, 243, V1394, 75, 332 (1564 and 1565), V1403, 56, 235; V1404, 28, 45, 72, 230.

4.7 Districts and Social Status

The district in which teachers lived may provide an indication of their social status. The 1579, 1584 and 1585 accounts of the Guild provide a full inventory of names of active teachers and, in most cases, their addresses. These lists show that most arithmetic teachers lived in the districts east and southeast of *Grote Markt.* On the basis of Scribani's 1610 map of Antwerp (see Figs. 4.3, 4.4, 4.5, 4.6), it is clear that there was a high concentration of arithmetic schools along the axis formed by Lange Nieuwstraat, Kipdorp and Katelijnevest/Minderbroedersrui. Most of the arithmetic teachers lived in better-off, though not exactly rich neighbourhoods. As was to be expected, few arithmetic teachers lived in the district around Grote Markt. The only exceptions are François van Loenhout and Melchior van Elselaer, whose homes were located along Oude Beurs. The subdistrict formed by Oude Beurs, Pandstraat and Doornikstraat,[69] three relatively short streets, yielded a revenue of 13,600 gl. in the second war tax, which is more than the whole of the third district. In other words, the latter two arithmetic teachers-cum-merchants lived at the heart of the wealthiest district. Melchior van Elselaer was one of the two teachers who were taxed 4 gl. in

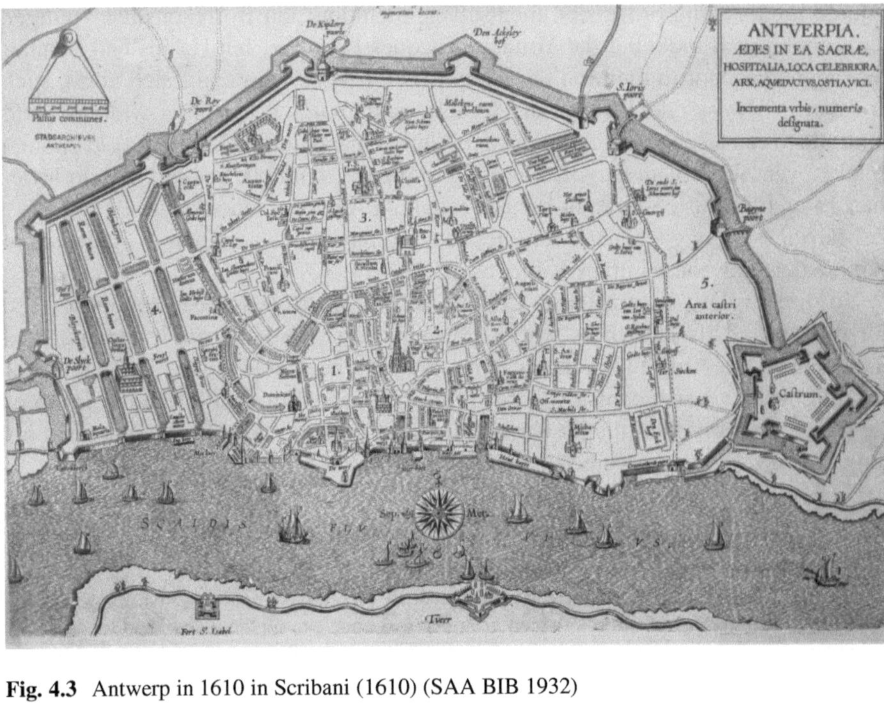

Fig. 4.3 Antwerp in 1610 in Scribani (1610) (SAA BIB 1932)

[69] Each district was subdivided into sub-districts or companies. The division was such that the civil guard company from each sub-district had about the same strength.

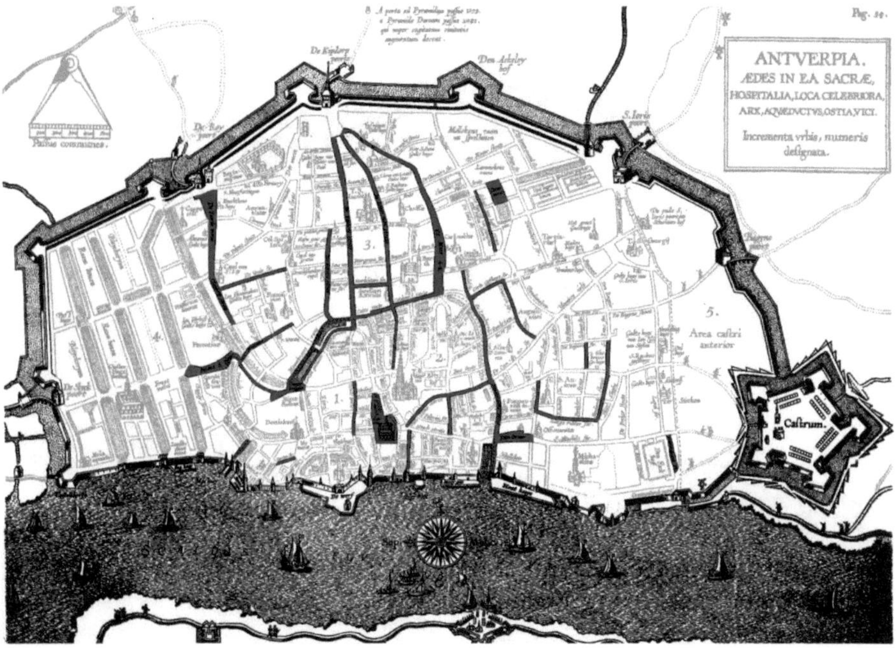

Fig. 4.4 Geographical spread in 1579

the monthly war tax, the highest amount among teachers,[70] so he must have been relatively wealthy. Michiel Coignet lived in a house on the corner of Braderijstraat and Grote Markt, the central market square, indicating that he, too, must by then have enjoyed a certain level of affluence.

Just a few arithmetic teachers, namely Peter Heyns, his son-in-law Kerstiaan Offermans, and Hans Lemmens, lived in a poor district. Heyns, however, was quite wealthy, as he was among the relatively small group of arithmetic teachers to be taxed in the monthly war tax, despite what the location of his home would suggest.

Disregarding administrative boundaries, we find that most teachers lived in areas near the commercial centre. Meir, Lange Nieuwstraat, Huidevetterstraat and Lange Gasthuisstraat were all located in the proximity of the Bourse. And Katelijnestraat and Minderbroedersrui were situated near the Bourse as well as Grote Markt. A smaller concentration of teachers lived in the 5th and 12th districts, near *Stadswaag*.[71]

[70] van Roey (1963, footnote 408).

[71] Goods weighing in excess of 50 pounds (about 23.5 kg) could not be sold before having been weighed at *Stadswaag*, for which a *waagrecht* or weighing tax was due. This *waagrecht* was payable each time the goods changed hands. Soly (1977, p. 133).

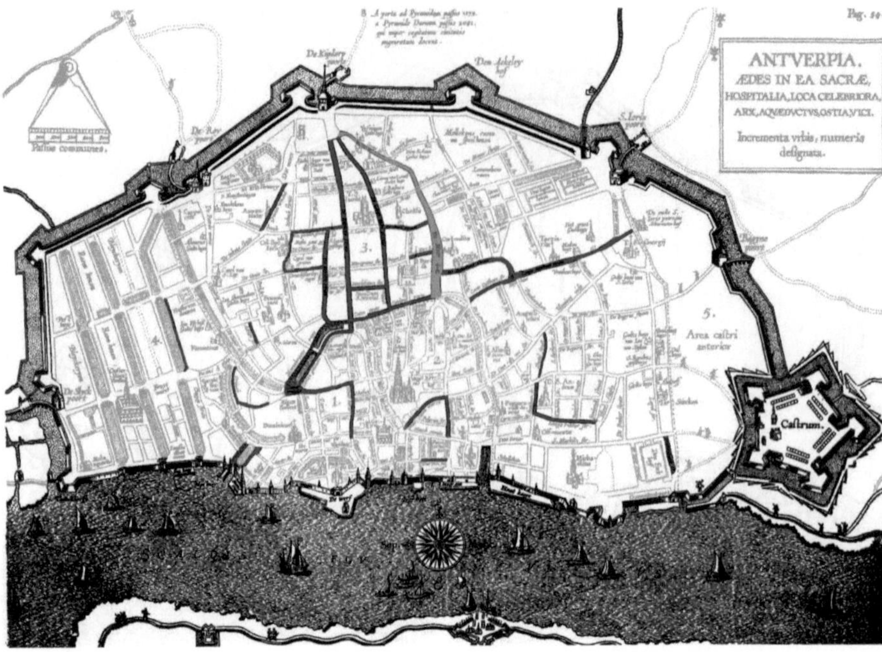

Fig. 4.5 Geographical spread in 1584

The presence of foreign enclaves may have influenced a teacher in his choice of residence. For instance, Hendrik Verhulst and Jan Godardus, who taught not only French but also another foreign language, lived respectively at Lange Gasthuisstraat and Huidevetterstraat, near the Portuguese quarter.

From the maps, it is immediately clear that many arithmetic teachers left the city during the religious troubles. By 1586, there were hardly any arithmetic schools in the western part of the city, unlike in 1579.

4.8 Property

The inventories of 1584 are also an invaluable aid in determining the location of the properties of the arithmetic teachers. In October of that year, a list of all houses, their owners and occupiers was drawn up, providing much the same information as can be found in the lists of the fifth penny tax.[72] We were able to determine the addresses of 43 of the 51 arithmetic teachers in Antwerp, as well as the rent charge value (*rentewaarde*, see footnote 7 on page 10) of their homes – an indication of

[72]For a detailed list, see Meskens (1994b, p. 87).

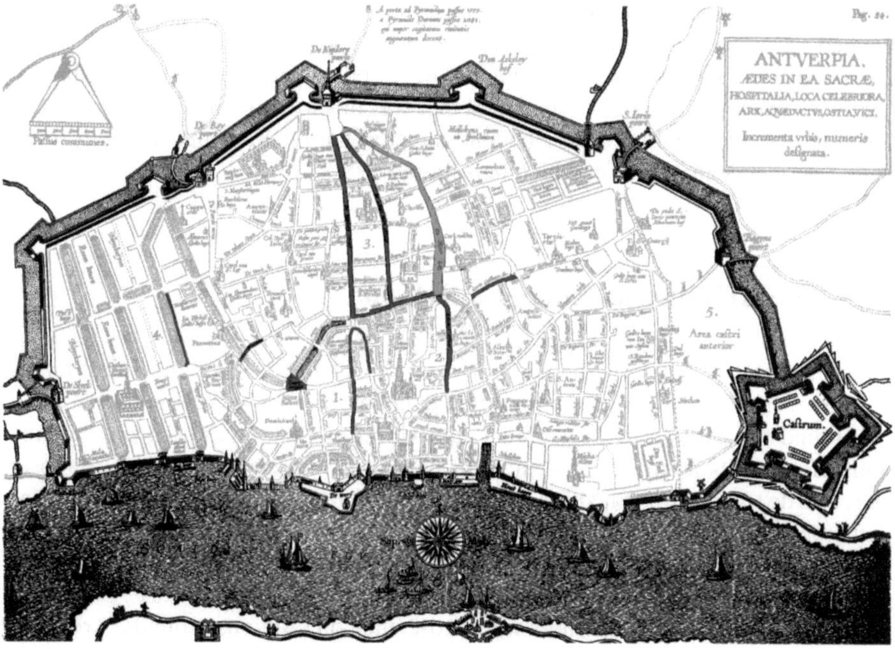

Fig. 4.6 Geographical spread in 1586

the value of the house they occupied. Most lived in rented accommodation; just 12 were owner-occupiers. The most expensive owned property was Anthony Smyters's *Het Gulden Calf,* estimated at a rent charge value of 214 gl. The cheapest owned dwelling was Hercules de Cordes's *De Penne*, at a rent charge value of only 45 gl.

The disparity among rented accommodation is even greater: three arithmetic teachers rented a house with a rent charge value over 300 gl., whereas Arnout de Cordes rented a home with a rent charge value of a mere 24 gl.

Rent value	Owners	Tenants
-50	2	6
50–99	3	11
100–149	1	9
150–300	3	6

The number of owners is obviously too small to draw any solid conclusions. Among the tenants, it is clear though that a majority of the schoolmasters occupied a modest or a middle-class home.

In the case of some arithmetic teachers, we were able to ascertain that they owned several properties. This information was obtained by chance rather than by systematic research, given the vast amount of archival materials potentially relating to real estate. Indeed, the properties concerned were spread across Antwerp and we had no prior indications regarding their locations. Peter Heyns owned at least six, relatively small, houses.[73] Michiel Coignet possessed at least three, which he agreed to sell to Jacques Cools in a deal that was thwarted by the untimely death of the latter.[74] Others held property beyond the city. Jan De Rademaker, for example, owned a tannery in Mechlin, which he sold in 1598.[75]

4.9 The Arithmetic Teacher's Religion

After the Sack of Antwerp, the number of teachers halved. Some may have left the city for economic reasons, but more often than not religion also came into play.

Around the middle of the century, the Guild closely monitored the religious obligations of its members and fines for non-observance were regularly imposed. Hercules de Cordes was fined for failing to attend Mass.[76] Valentin Mennher did not participate in the Offertory in 1550 and 1552 and was also fined.[77] As the century progressed, controls slackened. After 1560, no specific mention is made of this kind of offence.

Prior to 1566, just five teachers were prosecuted on religious grounds,[78] resulting in one execution. The events of 1566 provided an excellent opportunity to purge the Guild of heretics. The Spanish feared that heretical teachers would imbue their pupils with unorthodox ideas, which they might put into practice once they had risen to positions of power.

On 3 August 1568, some teachers were prosecuted for teaching their pupils the Beggar's Catechism and Beggar's psalms.[79] No fewer than 20 schoolmasters and 13 schoolmistresses were ordered to close their school.[80] In the aftermath of these events, many – including Peter Heyns – fled to Germany, to towns such as Cologne and Frankfurt, which already accommodated rather large Netherlandish colonies.[81] None, as it turned out, had left Antwerp for good, though: they all returned some years later, following the Pardon. Be that as it may, their temporary stay in Germany

---

[73] SAA GA4833, f371*v* and 377*v*, SR397, f160*r*-161*v*.

[74] SAA Proc. Suppl. 1739.

[75] SAA N466, f158*r*.

[76] SAA GA4528, f86*r*.

[77] SAA GA4528, f78*v*, f86*r*.

[78] Marnef (1996b, p. 63).

[79] Marnef (1994), Antwerpsch Archievenblad 12, p. 298.

[80] Bourland (1951, pp. 35–37).

[81] De Groote (1967a, p. 247), Marnef (1994, no. 272) and van Roosbroeck (1968, p. 183).

certainly seems to have resulted in closer contacts between arithmetic teachers from the two regions, as evidenced by their busy correspondence. It was in this period that Michiel Coignet entered the Guild.[82]

In a list of teachers of 1576, some names are preceded by the letter O.[83] This sign was also used after the Reconciliation by other Guilds and meant that the person in question was a Protestant (either Lutheran, Calvinist or Baptist).[84] No such indication was added to Coignet's name, which suggests he was regarded at this time to have been a Catholic.[85] However, in November 1585, on the lists of the Civil Guard, he was identified as a Protestant (either Calvinist or Anabaptist),[86] as was his assistant Jan de Rademaker,[87] and his brother Jacob was indicated as a Martinist (i.e. a Lutheran). These lists had been drawn up to form a Civil Guard consisting exclusively of Catholics and reconciled Protestants (i.e. those who intended to return to the "genuine Catholic Faith"). In principle, the Civil Guard consisted of all able bodied men aged between 18 and 60.[88] In practice, Calvinists and Anabaptists were excluded from the Guard and the armed Guilds. If Michiel had indeed been a Protestant previously, he was now deemed to be willing to reconcile himself with Catholicism.

Doubts concerning his religious conviction remain though. He is known not to have taken the Protestant oath in 1584, but it is unclear whether he refused to take it or whether he had given up his school. Moreover, by November 1585, Michiel was already a member of *Kolveniersgilde*, one of the four armed Guilds of the city.[89] This particular guild had already been "religiously cleansed".[90] This does not mean that all members were faithful Catholics, but it does indicate that they were at least willing to subscribe to the "true Catholic Faith".[91]

As for Michiel's brother, the painter Gillis, he decided after a couple of months' hesitation to leave the city because of his Protestant faith.

With the exception of some zealots on either side, it is very difficult to determine with any certainty to which religion individuals adhered. Very few teachers were willing to express their Catholic conviction during the period of the Religious Peace. One exception was Melchior van Aelst, who closed his school in order "to keep his faith". He had been a teacher at the school of Our Lady up to 1580.[92] That is not to say, though, that all teachers during this period were adamant Protestants. In fact,

---

[82]SAA Pk2933.

[83]SAA GA4550, pp. 330–331.

[84]See van Roey (1985a, pp. 203–204).

[85]For a full list of the religious convictions of the teachers, see Appendix A.3 on page 216.

[86]SAA GA4830(1), f52*v*.

[87]SAA GA4830(1), f366*v*.

[88]Boumans (1952a, pp. 743–750).

[89]SAA GA4663, ff. 321*v*, 325*r*, 344*v*, GA4664, f3*v*.

[90]Boumans (1965, p. 746).

[91]van Roey (1985a, p. 203).

[92]SAA GA4529, f93*v*, Prims (1951a, p. 215).

just a very few can be identified with any certainty as fervent Protestants, including Karel Strytberger, who taught at the Lutheran school of St George's, and Anthony Smyters, rector of the Calvinized school of St Andrew's.[93]

Peter Heyns[94] was a member of the Broad Council in the early 1580s. He was among the delegation that was sent by the city authorities to Prince William of Orange in Amsterdam for talks on 26–28 March 1581.[95] As clear an indication as any of where Heyns's sympathies lay. Later, after the Reconciliation, the minutes of a letter to the City Council (1588–1589) state that he was a Protestant.[96] Indeed, in a letter from Burgomaster Philips of Marnix, Lord of St Aldegonde, believed to be addressed to his negotiator, Richardot, Peter Heyns's name figures in a list of persons who wanted to leave Antwerp and for whom passports were to be requested.[97]

After the Reconciliation, the magistrate ordered the epuration of all Guilds. On 10 September 1585, all schools were closed down pending the outcome of this operation. On that same day, Aldermen Balthazar van Vlierden and Philips Dayala were appointed to look into the "competence of teachers",[98] which was clearly a euphemism for a process of religious cleansing whereby only Catholics were retained as Guild members. On 12 November, the *scholaster* approved the licences of 37 teachers, half the number there had been prior to 1585.[99] Among those who were forced to close down their schools was one Kerstiaan Offermans, Peter Heyns's son-in-law.[100]

It is clear that those who left immediately after the Reconciliation were either not Catholic or had collaborated to such an extent with the Protestants that they deemed it safer to flee. This is the case for Michiel Six and Jacob van Houthuys, Deans of the Guild of St Ambrose from 4 April 1585 up to the Reconciliation, and Peter Heyns, who had served as a *wijkmeester* (warden). All three men paid their debts to the Guild and left the city. The situation in other Guilds was similar. Michiel Coignet's brother Gillis had become Dean of the Guild of St Luke in 1584, a post he would continue to hold during and shortly after the Spanish siege. He subsequently fled the city in the Spring or the Summer of 1586.[101]

---

[93]Prims (1941a, p. 268), Prims (1951a, p. 216) and (Briels, 1980, p. 652).

[94]See Meeus (2000, 2009).

[95](Gielens, 1933, p. 225).

[96]Boumans (1952a), with a transcript of letters from SAA Pk278, dd. 10 February 1588. Also van Roey (1963, footnote 403) and Meskens (1998–1999). Ortelius was considered suspect because of his friendship with Peter Heyns (see chapter 9, p. 168).

[97]Gerlo (1988, pp. 51–52).

[98]SAA Pk558, f2*v*.

[99]For a complete list, see De Groote (1967a, 1968d), esp. p. 44ff for a list of schoolmasters persecuted for religious reasons.

[100]SAA Pk558, f3*v*.

[101]Rombouts and van Lerius (1874, I, pp. 315–316).

Being suspected of Protestantism could have grave consequences in later years as well. In 1588, Arnout Gillis complained that his wife was said to have been "contaminated with heresy", which almost led to the closure of his school.[102]

After the Reconciliation, Antwerp's population halved. The Deans of the Teachers' Guild requested the *scholaster* to limit the number of male and female teachers to 40, a request that was granted on 4 April 1588. The move had no direct consequences, as the number of teachers had in any case dwindled from around 80 to 37. If anything, the request seems to have been a strategy to protect the remaining teachers' livelihoods.

Evidence provided by the maps (see Figs. 4.3, 4.4, 4.5, 4.6) suggests that a massive emigration took place. Although the Sack of Antwerp undoubtedly prompted many to leave, this emigration trend had started much earlier. Previous studies concerning the resettlement of South Netherlandish teachers in the Republic between 1570 and 1630[103] have shown that it started slowly in the period 1570–1575, after which it gradually gained momentum until levelling off around 1585, to continue at a steady rate until ca. 1610.[104]

The first impetus for emigration was the imposition of the *Council of Troubles* following the Iconoclastic Fury of 1567. Some arithmetic teachers were banished, as indeed were many other citizens.[105]

After 1585 some were convinced they would be able to return in due course, as had been the case in the 1570s. Peter Keppens gave his father-in-law permission to take care of his possessions during his absence, clearly believing that he would not be gone for long.[106] Others were less confident of returning, and seemed to recognize that the present situation was quite different.

When Peter Heyns left Antwerp, he must have had an inkling that he would never return, as suggested by the fact that, in 1585, he paid his dues to the Guild. Previously, in 1567, he had left the city for Cologne, only to return in 1570.[107] Now he fled immediately after the Reconciliation, leaving his wife Anna Smits and his son Zacharias behind in Antwerp. Anna followed her husband soon after, joining him in Frankfurt. As for Zacharias, he may have begun his apprenticeship with Plantin at that time.[108] Having resettled in Frankfurt, Heyns opened a school for

---

[102] SAA Pk667, f174*v*.

[103] Briels (1972b, p. 97). The seminal work on immigration into what would come to be known as the Republic was published by Briels in a series of articles.

[104] Of the 418 immigrants who are known to have worked as teachers in the period 1570–1630, about a quarter (116) came from Antwerp. They settled mostly in Amsterdam (22), Delft (18) and Middelburg (18), and also in Leiden (13) and Dordrecht (11). Only 14 of these schoolmasters had previously taught in Antwerp. Briels (1972b, p. 98).

[105] Genard (1877a, pp. 58–59), Marnef (1994) and van Roosbroeck (1968, p. 138).

[106] SAA Cert47, f66*r*.

[107] Sabbe (s.d., p. 141), MPM, M394, ff1*v*-2*v*. Meeus (2000, 2009).

[108] Meeus (1988–1989, p. 600).

which he ordered books from Plantin.[109] In the first half of 1590, he moved his school to Stade, a town near Bremen, where he was entrusted with the religious instruction of girls.[110] The reason for this subsequent move is unknown. Perhaps he was a Calvinist, so that he may have run into difficulty in Lutheran Frankfurt. Only in 1592 did he agree to sell his properties in Antwerp.[111] Perhaps he had until then retained a glimmer of hope of one day returning to Antwerp, possibly after a victory by the troops of the States. Or perhaps it took until the 1590s for real estate prices to recover. After all, in a city that had lost about half of its population, property prices must have plummeted.[112] In 1594, Heyns left Stade and settled in Haarlem.[113] When a whale washed ashore in Berckhey (a fishing village on the coast of Holland) in February 1598, Peter Heyns rode out on a cart to witness the scene. He fell ill and died shortly after.[114] His granddaughter Catharina (daughter of Zacharias) married Johannes Bartjens, son of Willem Bartjens, who is regarded as the father of arithmetic teaching in the Netherlands.[115]

After the Sack of Antwerp, there was a quiet and modest reverse migration. Some arithmetic teachers were readmitted to the Guild, sometimes upon stating explicitly why they had temporarily left the city. Jacques Muytinckx had abandoned his school during the troubles and fled to Spain.[116] Robrecht van Huesden had moved to Zaltbommel, but subsequently decided to return.[117] Jan de Rademaker (Coignet's assistant teacher) and Tobias Stevens probably never left the city, but may have closed shop, as they were apparently readmitted in 1586.[118] All the teachers concerned may be assumed to have been Catholics who stopped teaching while the city was in Protestant hands, with a city council demanding a Protestant oath.

The Reconciliation and subsequent emigration turned the remaining teachers from producers of mathematical books into consumers.[119] The city never returned to the level of activity of pre-Reconciliation days, as Amsterdam gladly took over its leading role.

---

109 MPM Archief 63, Journal 1586, 13*v*, 40*v*, 118*v*.

110 Sillem (1883, pp. 571–572); on immigrants in Stade, see also Sillem (1893).

111 Briels (1972b, p. 103).

112 On this subject, see Soly (1974).

113 Meeus (2000) suggests that he followed the migration of Antwerp merchants, and Antonio Anselmo in particular (see p. 191).

114 Meskens (1996b, pp. 341–342).

115 Lambour (2004, p. 95). There appear to have been more connections between the Heyns and Bartjens families. Zacharias and Willem were godfathers to each other's children.

116 SAA GA4529, f129*r*.

117 Briels (1973, p. 123).

118 SAA GA4529, f116*v*.

119 For an overview of mathematics books printed in Antwerp, see Meskens (1994b, pp. 220–230). In the period 1550–1585, no fewer than 35 first editions of arithmetic books were published in Antwerp, as compared to ten in the period 1585–1620 (Meskens 1994b, p. 205).

Chapter 5
The Antwerp Arithmetic Books

Arithmetic books, more often than not written by arithmetic teachers, constitute a specific genre within the corpus of mathematics books. The content of these books is quite stereotypical, not only within the Netherlands but across the Western European trade zone. The way elementary arithmetic is taught at schools today can still be traced back to these books and their Italian manuscript forerunners.

During the Middle Ages, money consisted of silver or gold coinage. Travelling merchants would have to carry their cash along in a coffer, but with the likelihood of highwaymen lying in ambush, every journey was a hazardous undertaking. The bill of exchange, whereby a merchant could leave his money with a banker in, say, Venice and recover it in, for example, Bruges changed all that and made international trade a lot safer. From the thirteenth century onwards, European trade boomed. This increased the need for currency conversion, which is where the abacus teacher enters the story. The mathematics required for the valuation and conversion of currencies would ultimately lead to problems involving quadratic and cubic equations. Methods for solving such problems, of different levels of complexity, are dealt with in arithmetic books of the time.[1]

These books invariably begin with the pronunciation of numerals, the operations on integers and fractions, the rule of three and other practical rules. If the calculation of square roots is included, then also quadratic, rational and irrational equations. Some books deal with (arithmetic and geometric) series, ratios and cube roots. Just two deal specifically with wine gauging.[2] There are also a couple of books dealing with counting with jetons.

---

[1]For an analysis of the work of the abacus teachers, see van Egmond (1976); on the German *Rechenmeister*, see the studies edited by R. Gebhardt (Gebhardt and Albrecht 1996; Gebhardt 2002, 2005, 2008, 2011a); on Netherlandish *rekenmeesters*, see Kool (1999) and De Groote (1967a, 1968d).

[2]Raets and Coignet (1580b, 1597b) and vanden Dijcke (1600).

A. Meskens, *Practical Mathematics in a Commercial Metropolis: Mathematical Life in Late 16th Century Antwerp*, Archimedes 31, DOI 10.1007/978-94-007-5721-9_5,

© Springer Science+Business Media Dordrecht 2013

Fig. 5.1 Frontispieces of Mennher (1563) (*left*, MPM R50.29) and Mennher (1565) (*right*, MPM A 3589)

As for their use, a distinction can be made between different target audiences. The arithmetics by Gemma Frisius (1540 and numerous reprints[3]) were used first and foremost in Latin schools. Translations into the vernacular, however, indicate a need for good arithmetic manuals for non-academically trained readers. Handbooks aimed specifically at a mercantile audience were written by Ympyn, Mennher (see Fig. 5.1), Coignet and Vanden Dycke. Heyns's book was intended for craftsmen.

Remarkably, the only example of a geometry manual written by an arithmetic teacher is Mennher's *Traité des triangles spheriques* (1564). While the known arithmetic books do commonly contain problems involving geometry, these can generally be solved arithmetically.

The supremacy of Antwerp arithmetic books within the Netherlandish corpus is overwhelming. No fewer than 68 of the 109 editions (or 62 %) of sixteenth-century Netherlandish arithmetic books were published in Antwerp.[4]

[3]Reich (2005) has counted 101 sixteenth-century editions and another 64 seventeenth-century ones. These editions of the *Arithmetica* were printed in 18 different cities, and they include Latin, French, Italian, German and English versions, but remarkably not a single Dutch edition.

[4]See Smeur (1960a) and De Groote (1971).

Michiel Coignet was one of the foremost members of the publishing community, as becomes clear from his first publication. In 1573, he published an adaptation of Valentin Mennher's *Livre d'Arithmétique* (first published in 1561). The structure of this arithmetic book is classical. The first part deals consecutively with the four basic operations, the rule of three and the manipulation of fractions. After this introduction, it provides a number of exercises with facit (i.e. the numerical solution). The 1573 edition has the same basic content, but additionally it pays some attention to quadratic and third degree equations.

Usually, *Cent Questions Ingénieuses* was sold together with *Livre d'Arithmétique*. The former book was dedicated to Balthazar Schetz, Lord of Hoboken and one of Antwerp's wealthiest merchants.[5]

Coignet's book contains solutions to the 100 questions posed by Mennher in his 1561 book. Coignet occasionally changed the wording in order to clarify the question. The first 47 problems deal with arithmetic, the following 53 with trigonometry, often with relevance to solar or polar altitude. In these problems, Coignet demonstrates his ability to manipulate trigonometrical formulas. He appears to have been familiar with the trigonometric work of Regiomontanus and Copernicus. As a matter of fact, his erudition and knowledge of the mathematical literature would gain him respect throughout his career. Some bibliographies mention a book on accounting that he supposedly published in 1573, but no copy has hitherto been found, so that its existence is questionable.

In 1620 Denis Henrion published *Deux cens questions ingénieuses* in Paris. A year later, this work was republished as part of *Collection ou recueil de divers traictez mathematiques*. *Deux cens questions ingénieuses* consists of the 100 problems with their solutions that had previously been published by Coignet (1573) in *Cent Questions Ingenieuses*.[6] Henrion added to his book a further 100 problems taken from the work of Mennher. Henrion probably learned about Mennher's and Coignet's books during his stay in the Low Countries as an engineer with the troops of Prince Maurice. His aforementioned *Collection* also included a treatise on fortification in the Netherlands. As a matter of fact, Henrion never published any original mathematical work. Instead, he drew extensively from books written

[5]Wijnroks (2003, pp. 82–85). Balthazar Schetz was one of four brothers, whose business undertakings made them among the wealthiest people in Antwerp. Their father, Erasmus Schetz, had inherited a successful company that exported spices to Germany and he also owned a zinc mine in Kelmis (zinc is a constituent of brass). His wealth was so great that he was able to lend large sums of money to Charles V. His sons Gaspar, Melchior, Balthazar and Coenraet would later separate the banking and the trading activities into different companies. In 1555, Melchior became *buitenburgemeester*, giving him ample opportunity to mix public and private affairs, which he unashamedly did, to add even further to the wealth of his family.

When the Russian market was opened up, the Schetzes were among the first to jump at the opportunity, a move that yielded phenomenal profits. When their involvement in corruption became too obvious, they let their youngest brother's company default, leaving a debt of no less than 75,0000 gl! This did not however prevent them from continuing to trade and accumulate wealth.

[6]In *Cent Questions Ingenieuses* Coignet, in turn, gives the solutions to a hundred problems posed by Mennher in *Livre d'arithmetique* (1561), see De Groote (1970a).

by others authors and published under his own name. In 1616, he claimed to have invented the sector and even accused English author Gunter of having stolen his idea[7] (on the development of the sector, see Sect. 7.2).

During the 1570s, Michiel Coignet seems to have been very productive. In 1580, he reviewed and corrected Willem Raets's *Arithmetica* and his treatise *Vande Wisselroede* ("On the change rod" – a work on wine gauging, see Chap. 6). Little is known about Willem Raets apart from the fact that he was a merchant from Maastricht.[8] He was certainly not a teacher in Antwerp,[9] but does seem to have worked as a wine gauger in Maastricht.[10] This ties in with the fact that he twice entered the contest for the appointment of a new gauger in Antwerp.[11] He probably died between 1570 and 1576.[12] He seems to have maintained friendly relations with Michiel Coignet who calls him "our good friend in this and other arts of mathematics".[13] The first edition of Raets's *Arithmetica* was published in 1567 by Govaert Hamels and printed by Gillis Coppens van Diest.[14] Govaert Hamels was a teacher, who was admitted to the Guild in 1566–1567. By 1568 he had already left the city and, in his absence, was banished for his Lutheran convictions.[15] Gillis Coppens van Diest was the best mathematics printer of the day. In 1533, he became a member of the Guild of St Luke. Suspected of having printed Protestant literature, he was jailed for some time. Gillis Coppens van Diest, at the behest of Plantin, printed the first edition of Ortelius's *Theatrum Orbis Terrarum* (1570). The second edition of Raets's *Arithmetica*, as compiled by Michiel Coignet, was printed in 1580 by Hendrik Hendriksen. It would be reprinted in 1597 by Hieronimus Verdussen. This work, too, follows the classical pattern of a sixteenth-century arithmetic book. The four basic operations are explained, followed by the rule of three and 350 numbered problems. In addition, some attention is paid to series and square roots.[16] The treatise on wine gauging was the first Dutch book to mention the change rod (see Chap. 6, p. 106).

---

[7]Itard (1976, p. 271).

[8]SAA Pk639-640, f97*r-v* and 99r. Also Meskens (1993a, p. 26).

[9]His name does not appear in the list of teachers drawn up by De Groote (1967a, 1968d). Also Meskens (1993a, p. 26).

[10]Smeur (1960a, p. 36, footnote 3).

[11]SAA Pk1409, f15*r*–17*v*. Also Meskens (1994a, pp. 146–151). One of this contests took place in 1567, the other was presumably held between 1564 and 1572.

[12]The most recent archival source in which he is known to be mentioned is SAA Cert30, f251*v* dd. 1570. On the other hand, the privilege for his book was granted in 1576. This leads us to surmise that he died between these dates and most probably after applying for the privilege. In the preface to *Vande Wisselroede* (fMij*r*.), Coignet asserts "...de welcke (door sijne subite doot) hieraff beledt is gheweest", (which [= making his book print-ready] his sudden death prevented him from doing).

[13]Raets and Coignet (1580a), fA iij*r* and Mij*r*.

[14]Smeur (1960a, p. 36) and Smeur (1960b).

[15](De Groote, 1967a, p. 262).

[16]Smeur (1960a, pp. 62–63).

The manuscript *Livre d'Arithmétique* (1587) is a revision of the 1573 Mennher-Coignet arithmetic book (Fig. 5.6 on p. 68). It was written in 1586–1587, but contains later additions. While the manuscript is based on the 1573 book, its purpose is unclear. Was this a manual for a student, providing the solutions to the problems, or was it the manuscript for a new edition of the book? The latter cannot be ruled out. Even in 1603, John Hay s.j. urged Clavius to publish Coignet's arithmetic book,[17] hoping that he would find it a valuable addition to his proofs of Euclid's tenth book of the *Elements*.

In his manuscript, Coignet gives solutions to all the problems he proposes. The vast majority of these problems also appear in the 1573 book. Some have been omitted, and occasionally he adds a new example. The manuscript is superior to the book in that its exposé is clearer.

5.1 The Use of the Arithmetic Manuals

Considering that the first Latin manuscript to use Hindu-Arab numerals, the Spanish *Codex Vigilanus*, dates back to 976[18] and that Fibonacci's popular *Liber Abaci* was written in 1202, it took a long time for such numerals to become established in daily practice. In Antwerp, and presumably in the Low Countries in general, Hindu-Arab numerals were introduced in the sixteenth century.

The transition from Roman to Hindu-Arab numerals is documented in the accounts kept by merchants[19] and Guilds.[20] In Heyns's arithmetic manual *Tot Profyte* (1561), some examples are set in Roman numerals while others use the Hindu-Arab script (See Fig. 5.2). The alternation between the two systems is not random: the left pages are reserved for Roman numerals, the right ones for Hindu-Arab numerals. The book was presumably laid out in this way in order that Heyns could teach his girls the two existing systems simultaneously.

[17]Baldini and Napolitani (eds.) (1992, nr. 204).

[18]Struik (1980, p. 100).

[19]The journals of Frans de Pape, a cloth merchant and keeper of the wine excise, cover a lengthy period of time. He seems to have began making the switch from Roman to Hindu-Arab numerals around 1560. This transition was made at different times in different kinds of accounts. SAA IB482-496.

[20]Consider, for example, the account books of the Schoolmasters Guild. The first use of Hindu-Arab numerals was in 1545, under the deanship of Lodewijk Mindercourt and Jacob Huyssens (an arithmetic teacher). Five years later, in 1550, Roman numerals were reintroduced and Hindu-Arab numerals appeared only sporadically (1553, 1566, 1567, 1569 and 1578). By 1580, Hindu-Arab numerals seem to have won the day. Although the accounts of 1580, 1581, 1583 and 1584 contain some Roman numerals, in the running text preference is given to Hindu-Arab numerals. After 1585, Hindu-Arab numerals are used in all accounts save one (1600).

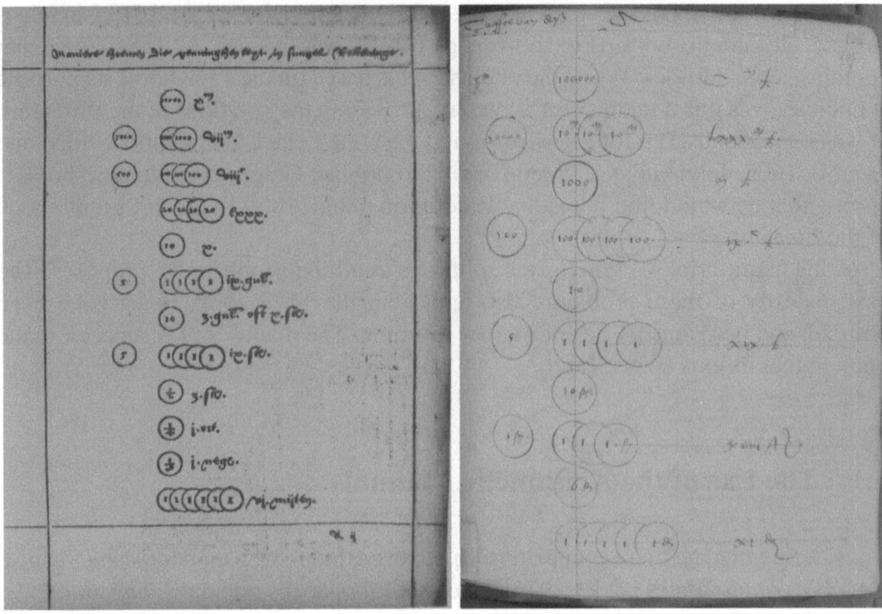

Fig. 5.2 Jetons in Peter Heyns (MPM BM31561) and Gerard Gramaye (SAA IB # 766)

Although Roman numerals do not appear in the arithmetic books, with the exception of books on the Casting Counter,[21] we should be careful not to draw premature conclusions regarding the acceptance of Hindu-Arab numerals. It is possible that books using Roman numerals, such as Van Huesden's *Rekenboeckxen* (1564), were actually widely distributed, but that they have not been preserved because of their limited value. J.B. De Vos edited Van Huesden's *Rekenboeckxen* as late as 1695, which makes it quite feasible that it was reprinted many times during the seventeenth century.

It would be very interesting to have insight into numeracy rates, particularly in a city such as Antwerp. However, it is already difficult to estimate the extent of literacy,[22] let alone that of numeracy. Nonetheless, with the exception of counting with jetons (Fig. 5.2), arithmetic requires some literacy, so that we know the degree of numeracy will have been lower than that of literacy.

We have no clear picture of who used which arithmetic manuals. There are no surviving school curricula to rely on, with the exception of those of the Latin schools, but these are not really representative for the subject at hand. Arithmetic training in these schools only started in the second year, and the most commonly used book was Frisius's *Arithmetica* (Fig. 5.3).

[21]On the Casting Counter, see Barnard (1981).

[22]Parker (1980), also Graff (1983).

ARITHMETICAE PRACTICAE METHODVS FACILIS, PER GEMMAM FRISIVM MEDICVM AC MATHEMATICVM.

Fig. 5.3 Detail of the frontispiece of Gemma Frisius's *Arithmeticae practicae methodus facilis,* 1540 (EHC G 55779)

Valcooch[23] recommended Robrecht van Huesden's book for basic arithmetic, in particular counting with jetons. For advanced arithmetic, Cresfelt's *Arithmetica* (1557) was deemed well suited.

It has been noted in relation to Ramus editions[24] that the format is an indication for their use: octavo and duodecimo for school use, quarto and folio for collectors and libraries. Most of the Antwerp arithmetic books have a small format (8° or 16°). Following the same logic, this may indicate that these books were used primarily in schools.

---

[23]De Planque (1926, p. 36).

[24]Maclean (1990, p. 261).

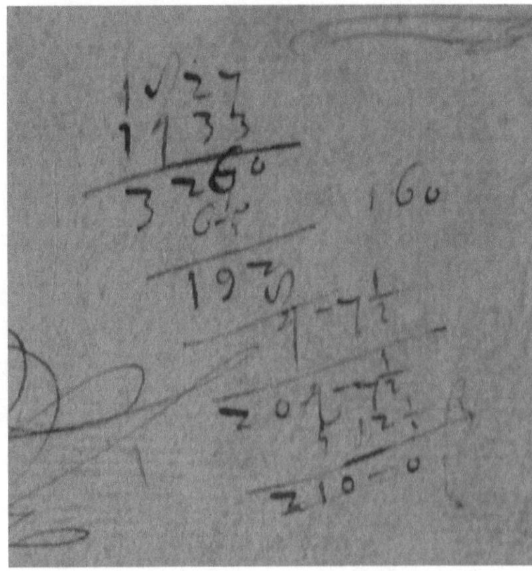

Fig. 5.4 Marginal notes in Serlio's *Architectura*. One recognizes $204-7\frac{1}{2}+5-12\frac{1}{2} = 209-20$ or 210 gl (EHC H 140382)

5.2 The Rule of Three

The basic arithmetical operations were known as "species".[25] The sixteenth century authors do not however agree on the number of species: was the pronunciation of figures a fifth species or not? And halving and doubling included, the number of species could add up to seven. The species were treated in all arithmetic manuals, and examples could include long and tedious calculations. Sporadically, three further species were introduced: adding, subtracting and duplicating with non-decimal units (as in money problems – see Fig. 5.4).[26]

The rule of three is a method for solving proportions. This can be done relatively easily using algebra. In brief, it says that if you know three numbers a, b, and c, and want to find x such that

$$\frac{x}{a} = \frac{b}{c} \text{ (that is, x:a::b:c) then } x = \frac{ab}{c}$$

It is a rule that was known throughout the world since ancient times[27] and, as we shall see (see p. 71), would continue to be used in some ingenious devices to calculate the value of certain non-linear functions.

[25]For a detailed analysis of the content of the arithmetic books, see Smeur (1960a), Meskens (1994b) and Kool (1999).

[26]E.g. in Nederlands Economisch Historisch Archief (Amsterdam) NO351; on this manuscript see Meskens (1993b).

[27]Shen (1999) and Tropfke (1980, p. 359).

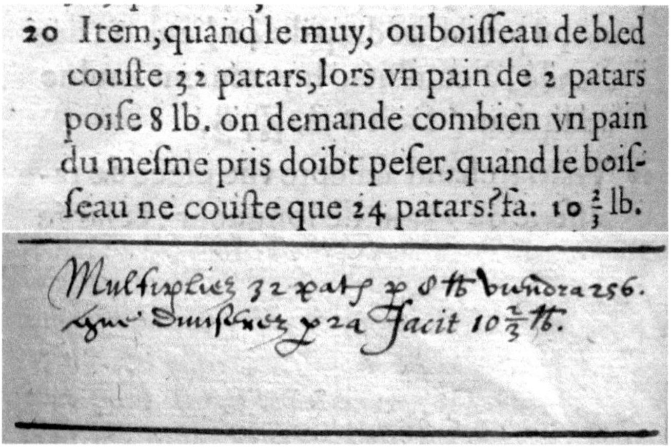
20 Item, quand le muy, ou boiſſeau de bled couſte 32 patars, lors vn pain de 2 patars poiſe 8 lb. on demande combien vn pain du meſme pris doibt peſer, quand le boiſſeau ne couſte que 24 patars? fa. 10 $\frac{2}{3}$ lb.

Fig. 5.5 A rule of three question and its solution, presumably by the owner of the book, on intershot white pages. From Mennher (1565) (MPM A 3589)

With the exception of the books on the casting counter, all Antwerp arithmetic books carry numerous examples of problems that are resolved using the rule of three. Some are devoted in their entirety to applications of this rule (Fig. 5.5).

Not without reason was this rule referred to as the Golden Rule by, for example, Peter Heyns (1561, p. 33):

> Den reghel van dryen is den bequaemsten, nutsten en orborlycksten onder alle reghelen/want hij is alle andere reghelen behulpsaem. Waeromme hij van vele te rechte ghenoemt wordt den gulden reghele/bewijsende daer mede zijn weerdicheyt. Een yeghelijck behoort wel neerstich te sijne om dien te leerene en te verstane.
>
> *The rule of three is the most useful of rules, because it is an aid to all others. It is rightly referred to by many as the "Golden Rule", which testifies to its value. Everyone should study it diligently to learn and understand it.*

Raets-Coignet's *Arithmetica* (f Cij*v*) explains the rule as:

> Desen Regel heeft vier ghetalen staende deen teghen den anderen in proportie ghelijck den eersten teghen den tweeden alsoo den derden teghen den vierden: soo volcht dan byder 14. vij element. dat alsulcke proportie alsser is tusschen deerste en tderde alsulcke proportie isser oock tusschen het tweede ende tvierde.
>
> [...]
>
> Dus soo volget by den 19. Vij element, alsmen multipliceert den eersten met den vierden daer soude soo veel comen als oftmen multipliceerde den tweeden inden derden ...
>
> *This Rule involves four numbers which are in proportion to one another: the first [number] is to the second [number] as the third [number] is to the fourth [number]. From the 14th [proposition] of Elements VII [= the seventh book of* Euclid's Elements*] it then follows that the first [number] is to the third [number] as the second [number] is to the fourth [number].*
>
> [...]
>
> *From the 19th [proposition] of Elements VII it follows that the result of the multiplication of the first by the third [numbers] is the same as the result of the multiplication of the second by the fourth [numbers].*

The first examples are concerned with the reduction of money:

Item eenen β gelt 12.ϑ hoeveel ϑ doen 16 β.
Item one β holds 12.ϑ how many ϑ will 16 β hold. (f *C*vr)

Applying the rule of three in a non-decimal system requires numerous examples of multiplication:

Item soo een marck sijns siluers cost β 45ϑ 6 wat costen 54 marcken 4 oncen
Item if a mark silver costs β 45 ϑ 6 what do 54 marks 4 ounces cost (f *D* viiir)

	M	oñ		M	oñ	
	54.	4.		54.	4.	
	45.	6.		2.	5.	6
	270.			108.		
	216 .		4β	10.	16.	
6ϑ	27.		1β	2.	14.	
4 oñ	22.	9.	6ϑ	1.	7.	
	2479.	9.	oñ	1.	2.	9
Facit	lb. 123.19.9		Facit	lb. 123.	19.	9

On the left hand, the multiplication is performed with stivers and then converted to pounds. On the right, the stivers are first converted into pounds and subsequently the multiplication is performed. To understand how this multiplication is carried out, one needs to know that £ $1 = \beta\, 20$ and $\beta\, 1 = \vartheta : 12$, while 1 mark silver holds 8 ounces.

$$\begin{aligned}
&54\text{M } 4\text{oñ} \times \beta\, 45\, \theta 6\\
&= \left(54 + \frac{4}{8}\right)\left(40 + 5 + \frac{6}{12}\right)\\
&= 54 \times 5 + 54 \times 40 + \frac{6}{12} \times 54 + \frac{4}{8}\left(45 + \frac{6}{12}\right)\\
&= 270 + 2160 + 27 + \frac{4}{8}\left(44 + \frac{18}{12}\right)\\
&= 270 + 2160 + 27 + 22 + \frac{9}{12}\\
&= 2479 + \frac{9}{12}
\end{aligned}$$

Shown on the right is a similar calculation, but here the stivers have first been converted into pounds. The calculation is slightly more complicated, but the method is the same.

$$
\begin{aligned}
&54\text{M } 4\text{oñ} \times £2\ \beta\ 5\ \theta\ 6 \\
&= \left(54 + \frac{4}{8}\right)\left(2 + \frac{4}{20} + \frac{1}{20} + \frac{6}{240}\right) \\
&= 54 \times 2 + 54 \times \frac{4}{20} + 54 \times \frac{1}{20} + 54 \times \frac{6}{240} + \frac{4}{8} \times \left(2 + \frac{5}{20} + \frac{6}{240}\right) \\
&= 108 + \left(10 + \frac{16}{20}\right) + \left(2 + \frac{14}{20}\right) + \left(1 + \frac{7}{20}\right) + \left(1 + \frac{2}{20} + \frac{9}{240}\right) \\
&= 123 + \frac{19}{20} + \frac{9}{240}
\end{aligned}
$$

Application of the rule of three inevitably leads to calculations involving fractions. These fractions can yield slightly odd results, especially in problems where realistic values lead to unrealistic solutions. For example, one is very unlikely to pay or receive £634 − 2 − $5\frac{3836876359554905088}{25000000000000000000}$ (Coignet (ca. 1587), f63*v*, see Fig. 5.6).

The rule of five, or the combined rule, is a special application of the rule of three

A	B	C
a_1	b_1	c_1
a_2	b_2	c_2

whence $b_1/a_1c_1 = b_2/a_2c_2$ or $a_1b_2c_1 = a_2b_1c_2$.

The *regula falsi* is a method for solving equations, leading to first-degree equations, without actually needing to formulate an equation. In modern terminology, the rule says that the root of $y = f(x) = ax + b$ is given by[28]

$$x = \frac{x_2 f(x_1) - x_1 f(x_2)}{f(x_1) - f(x_2)}$$

The rule was commonly applied and it appeared in the books of Frisius (1540), Vandenhoecke (1545), Mennher (1550, 1556, 1563, 1565), Raets (1580), and Mellema (1586).

[28]It is easy to see the validity of the rule, by calculating the eliminant of the system (*S*) in *a* and *b*.

$$(\text{S}) \begin{cases} ax + b = 0 \\ ax_1 + b = f(x_1) \\ ax_2 + b = f(x_2) \end{cases}$$

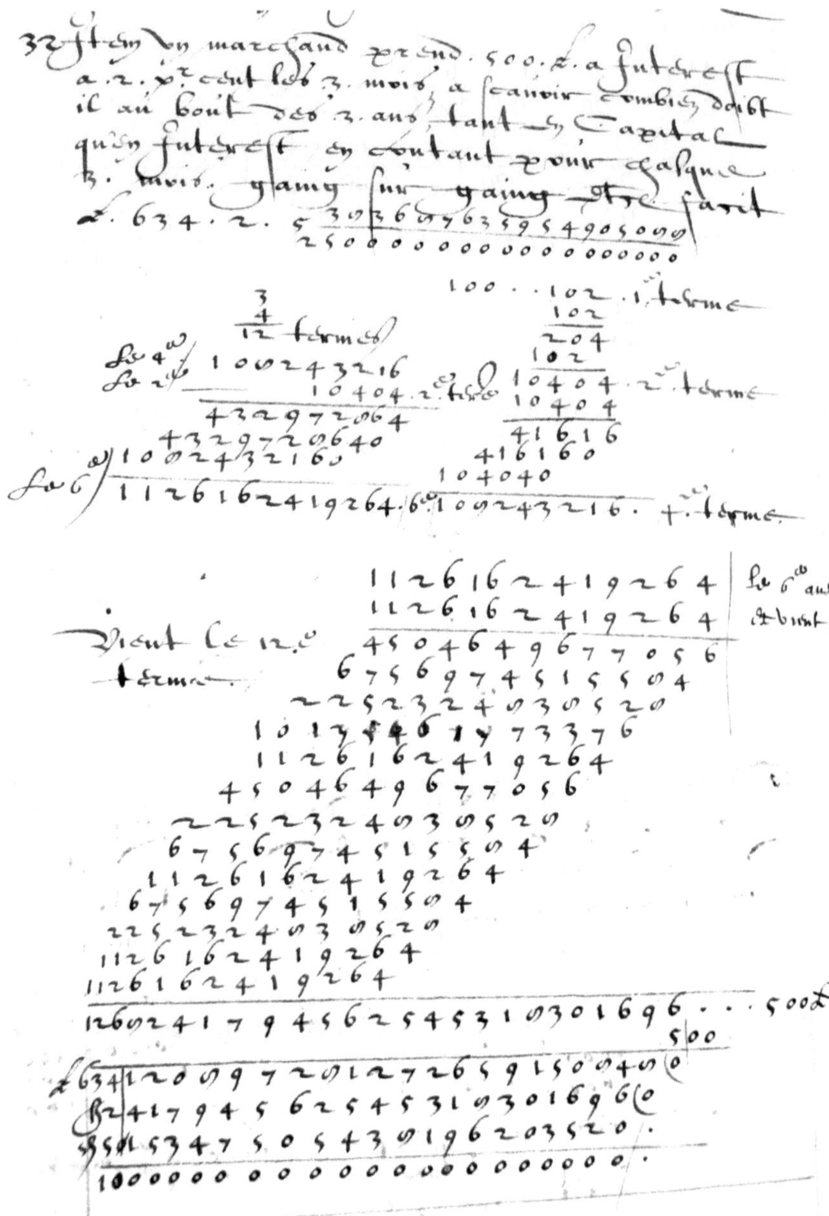

Fig. 5.6 Calculation of the final capital lent at 2 % per quarter for three years. First 1.02^{12} is (correctly) calculated, by repeated multiplications. This result is multiplied by 500 and expressed in pounds (Coignet ca. 1587, f63*v*, SAA IB#2964)

5.3 Barter

Among the most popular kinds of problems were those involving barter.

For instance:

Deux marchant barratêt leur marchandise l'un a l'autre/l'un ha drapz à £5 argent comptant et en troque il les mect a £5½ et l'autre ha velours à β 10 à combien le doibt il metre en troque qu'il soit egal.
10 1φ 5 faict ½ eg. à 5½
faict 1φ egal à β11

Two merchants wish to barter their goods. One merchant has cloth which is worth £5 silver, but which he wants to barter at a price of £5½. The other merchant has velvet which is worth β 10. At what price does he have to trade his velvet for the exchange to be fair:

10 1φ 5 makes ½ equal to 5½
makes 1φ equal to β11
(Mennher 1565, f Zvij *v*).

In a standard formulation, the problem amounts to a situation where two merchants A and B wish to barter goods, the prices A_1 and B_1 of which are known. A wants to trade at a higher price A_2, which gives rise to the problem of which price B_2 will result in a fair exchange for B (i.e. which price will ensure that the same amount of goods changes hands as with the original prices).

In other words, one has to determine B_2 from $A_1/B_1 = A_2/B_2$.

In other problems, one of the merchants wants to make a profit, leading to the equation:

$$m(A_1 + p) = nB_1$$

in which m and n are the amounts of goods A and B agree to barter and p the profit.

In some problems, one of the two parties wants to be paid for a part $(1/n)$ of the goods traded:

$$\frac{A_1 - \frac{B_1}{n}}{B_1} = \frac{A_2 - \frac{B_1}{n}}{B_2}$$

It is important to note that, in all cases, the barter is measured against a common abstract standard: money. Hence the deal is not driven by a need for the goods as such, but by a desire to own goods with which things can be acquired.

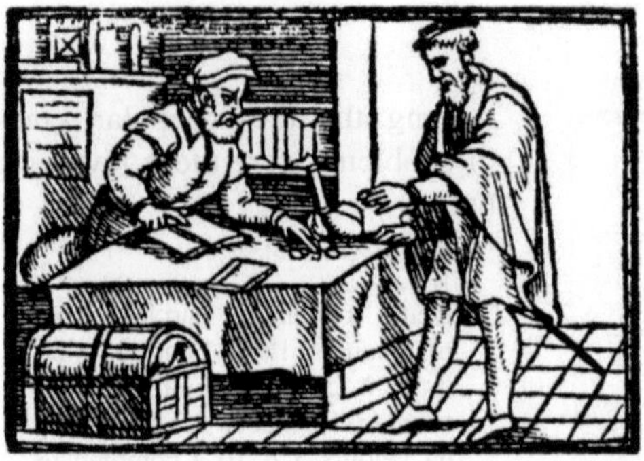

A particular kind of barter is that involving different currencies. It was the kind of problem that was self-evidently taught in a mercantile city such as Antwerp. Solving such queries required some familiarity with the subdivisions of the various currencies. Most problems were however quite straightforward and easily solved with the rule of three, e.g.

> Un marchant à Noremberg ha donne flo. 650 kreu. 35 pour les rauoir à preslau tousiours kr. 60 pour 28 gros polonois combie luy reuiendra à preslau à gros 30 par flo.
> 1..60..650..35 faict 39035
> 60..28..39035 fa fl. 607 gros 6 po.
> *A merchant in Nuremberg has given flo 650 kreu 35 [to a banker] with the intention of retrieving the money in Breslau [now Wrocław] at the same exchange rate. The exchange rate is kr 60 for 28 Polish gros. A Polish florin holds 30 gros. What amount will he receive in Polish money?*
> *1..60..650..35 makes 39035*
> *60..28..39035 makes fl. 607 gros 6 po.*
> (Mennher 1556, f Dij*r*)

This problem combines two rules of three. First, the florins are expressed as kreuzers (1 florin equalling 6 kreuzers). At Breslau, 60 kreuzers are valued at 28 Polish groats, so the merchant receives $39,035.\frac{28}{60} = 18,216\frac{1}{3}$ Polish groats for 39,035 kr. Since a Polish florin holds 30 groats, the solution is fl. 607 gros 6.

5.4 Algebra and the *Rule of Coss*

The fourteenth-century Italian abacists still used rhetoric algebra. Problems and their solutions were given in words, without abbreviations or symbols.[29] The unknown was referred to as *cosa*. This terminology was also used in Southern Germany, where it came to be known as *der Coss*.

Sixteenth-century arithmetic books show a diversity in symbolization, an indication that things were on the move. Symbols were attributed to the powers of the

[29] van Egmond (1976, pp. 217–222). On the emergence and development of symbolic systems, see Heeffer (2006a).

unknown. We can roughly distinguish between two schools, namely the Italian and the Southern German school. The symbolism of the latter would come to be used throughout Western Europe.

The earliest indications of proto-symbolic algebra are found in Italy. In his *Summa de aritmetica* (1494), Luca Pacioli uses the following abbreviations[30]:

x^0	x^1	x^2	x^3	x^4	x^5	x^6	x^7
n^o	co.	ce.	cu.	ce.ce.	p^o r^o	ce.cu	$2^o r^o$
numero	cosa	censo	cubo	censo de censo	primo relato	censo de cubo	secundo relato

This system was used until the first half of the sixteenth century, not only throughout Italy by, among others, Tartaglia (*Nova Scientia*, 1537, and other manuals) and Cardano (*Practica arithmeticae generalis*, 1539, 1545, 1570), but also by the Portuguese mathematician Pedro Nuñez.[31]

Regiomontanus was one of the progenitors of mathematical symbolism in the West. Already in his manuscripts of 1456, he used symbols for the unknown and its square, for the root, the subtraction and the equality. Whether this symbolism was original is doubtful. It is also used by the monk Fridericus Gerhardt (†1464/65) in a manuscript from 1461.[32] Therefore, one may assume it to have been more or less commonplace in Southern Germany.

x^0	x^1	x^2	x^3	x^4	x^5	x^6	x^7
ϑ or N	ꝛe	ʒ	ce	ʒʒ	ß	ʒꝛe	bß
Dragma or numerus	radix	zensus	cubus	zens de zens	surso-lidum	zensi-cubus	bisur-solidum

At first sight, the ꝛe resembles a sloppily written *res*,[33] while the ce seems to be a stenographic abbreviation of *cubus*.

In Western Europe, this notation gained quite a few followers, such as Michael Stifel (*Arithmetica Integra*, 1544, 1545, 1553), Valentin Mennher and Michiel Coignet (*Livre d'Arithmétique*, 1573), Robert Recorde (*Grounds of the Artes*, 1557), Jacques Peletier (*Algebra*, 1554) and many others.[34]

[30](Cajori, 1993, pp. 107–109).

[31]Cajori (1993, pp. 117–123 and 161–164).

[32](Folkerts, 1977, pp. 222–223), Folkerts (1996, pp. 24–27) and Schröder (1996, pp. 32–34).

[33]Latin for *the thing*. It is the same terminology as Italian *cosa*. See also Tropfke (1980, p. 377).

[34]Meskens (1994b, pp. 64–66) and Cajori (1993, pp. 139–147, 164–167, 172–177).

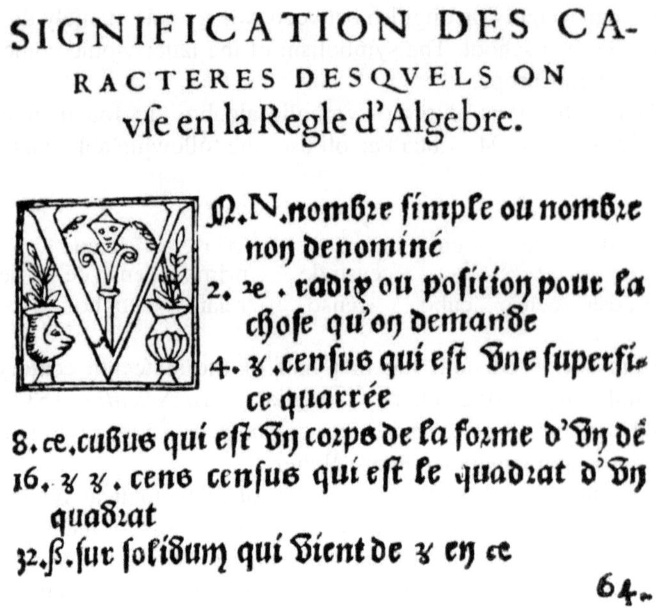
SIGNIFICATION DES CARACTERES DESQVELS ON vſe en la Regle d'Algebre.

VN.N.nombre ſimple ou nombre non denominé
2. ꝛe . radix ou poſition pour la choſe qu'on demande
4. ʒ .cenſus qui eſt vne ſuperfice quarrée
8. ce.cubus qui eſt vn corps de la forme d'vn dé
16. ʒ ʒ . cens cenſus qui eſt le quadrat d'vn quadrat
32. ß.ſur ſolidum qui vient de ʒ en ce
64.

Fig. 5.7 Cossic symbols (Mennher 1556, EHC G 21409)

In Mennher's, books we encounter ((1565, Piiij*r*), see Fig. 5.7):
Signification des Caractères, desquelles on use en la Règle d'Algèbre

	Vn N. signifie simple ou nombre non denominé
1.	ꝛe radix ou position pour la chose qu'on demande
4	ʒ census, qui est vne superfice quarré
8	ꝛe cubus, qui est vn corps de la form d'vn dé
16	ʒ ʒ cens census, qui est le quadrat d'vn quadrat
32	ß sur solidum, qui vient de ʒ en ce
64	ʒ ce cens cubus, qui vient de ʒ en ʒ ʒ
128	bß bissursolidum, qui vient de r en ʒ ʒ
256	ʒ ʒ ʒ cens cens census, qui vient de ʒ ʒ en ß
512	ce ce cubus de cubo, qui vient de ʒ ʒ en ß &c.

The names still refer to the Italian origin, but the symbols are German. The ß has to be read as a German *sz*.

To simplify matters, Mennher (1565) provided a table and rules for its use:

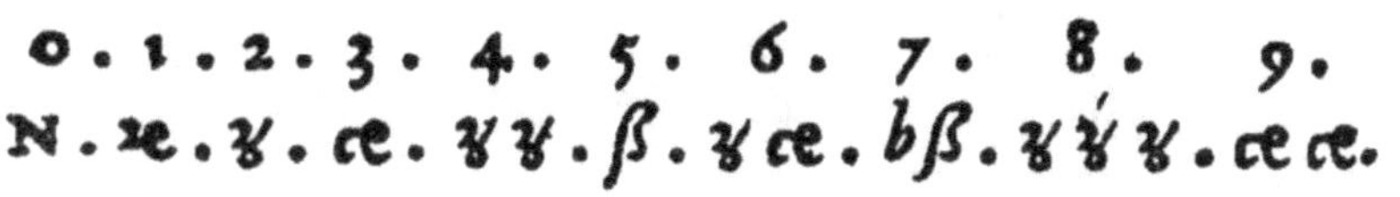

... si vous desires sçavoir qlle quantité il en sorte quand on multiplie 1℞ par 1ℨℨ regardes sur le quel ciffre il y a & vo trouueres un/et pareillement surℨℨ et trouueres 4 lesquels adioustes ensembles qui seront 5 et puis la mesme quantite qu'elle ha sur soy le 5 icelle en sortira quand on multiplie 1℞ par 1ℨ et c'est ℨ

... *if you wish to know which number one obtains by multiplying 1℞ by 1ℨℨ then look in the first row [of the table] for the number above ℞ and you will find 1; similarly above ℨℨ you will find 4. Now add these two numbers together, which gives you 5. [Look at the table once more and] in the second row under the 5 you will find β. Thus the result of the multiplication of 1℞ and 1ℨℨ is 1 β*

(Mennher 1565, f Qij *v*).

Michael Stifel proposed an alternative system, in which the unknown is represented by a letter. The number of consecutive letters equals the power, thus x^4 would be written as $1AAAA$. The idea of representing the unknown by a letter would, however, not catch on for a few more decades. Nevertheless, this notation sometimes co-existed with cossic notation. Mennher actually uses the two systems together in a number of problems.[35] Other systems use a symbol that we would refer to as an exponent. One author to apply such a notation was Heinrich Schreiber (also known as Grammateus):

x^0	x^1	x^2	x^3	x^4	x^5	x^6	x^7
N	Pri	Se.	3^a	4^a	5^a	6^a	7^a
Numerus	Prima	Secunda	Tercia	Quarta	Quinta	Sexta	Septima

It was also used by, among others, Gielis Vandenhoecke (*Een sonderlingh boeck*, 1545 – Fig. 5.8).[36] Again, what appears to be an exponent is actually an abbreviation of the name of the power.

In essence, the same system is applied by Rafael Bombelli (*L'algebra*, 1572, 1579),[37] though here a major step has already been made towards symbolic notation. Bombelli writes ax^n as $\overset{n}{\smile}\!\!\!\!a$.

[35]Mennher (1565, F f iij ff.).

[36]See Bockstaele (1985, esp. pp. 17 ff.).

[37]Cajori (1993, pp. 123–128).

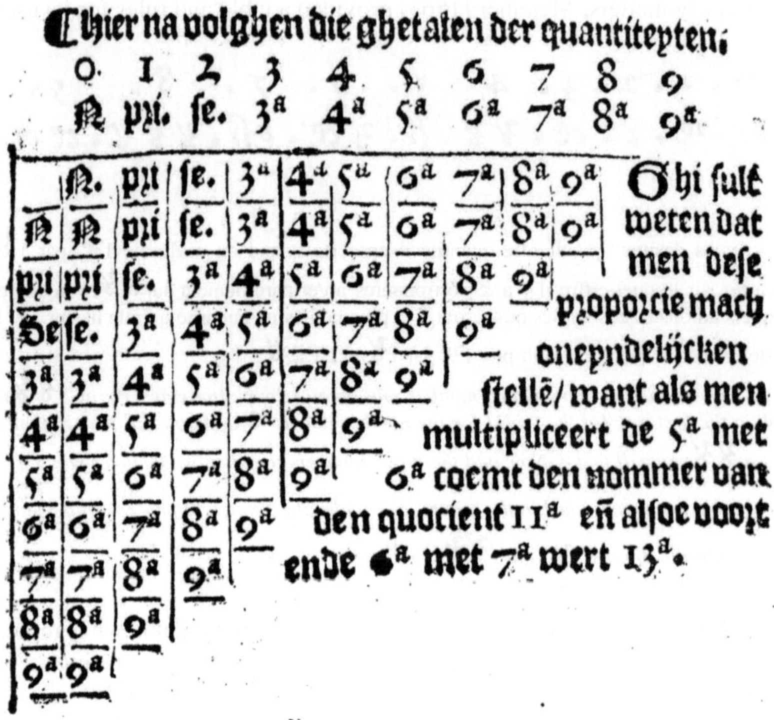

Fig. 5.8 Exponentiation in vanden Hoecke (1545) (MPM R 50.28)

A similar system is used by Stevin, who writes ax^n as a$\textcircled{n}$. Problematically, Stevin uses the same notation for decimal numbers. For instance 1$\textcircled{0}$ 3$\textcircled{1}$ 0$\textcircled{2}$ 2$\textcircled{3}$ stands for 1.302, while 3$\textcircled{1}$ + 2 means $3x + 2$, but could also be interpreted as 0.3 + 2 (= 2.3).

Viète uses letters for the unknown, but he continues to name the exponent verbatim or abbreviated. The names he uses for the powers are the Latin equivalents of Pacioli's names.

Thomas Harriot was the first to use small letters for the unknown (in Stifel's second notation). It is only with James Hume and René Descartes that the system of exponentiation that we use today came into general use.[38]

The polynomial $9x^5 - 7x^4 + 5x^2 - 3x + 1$ thus becomes

with Pacioli	9.p^{o}r^o.m.7.ce.ce.p.5ce.m.3.co.p.1.n^o
with the cossists	$9\beta - 7\mathfrak{zz} + 5\mathfrak{z} - 3\varphi + 1$
with Stifel	$9AAAAA - 7AAAA + 5AA - 3A + 1$

[38] Cajori (1993, pp. 188–209).

with Grammateus	$95^a - 74^a + 5se - 3pri + 1N$
with Stevin	9⑤– 7④+ 5②– 3①+ 1
with Viète	A plani cubo 9 – A planoplano 7
	or A pl.c.9 – A pl.pl.7 + A pl.5 – A3 + 1N

Using these notational systems, solutions to equations were presented. They were usually reduced to four known types :

$$ax = b \text{ (and } ax^n = b)$$
$$ax^2 + bx = c \text{ (and } ax^{n+2} + bx^{n+1} = cx^n)$$
$$ax^2 + c = bx \text{ (and } ax^{n+2} + cx^n = bx^{n+1})$$
$$ax^2 = bx + c \text{ (and } ax^{n+2} = bx^{n+1} + cx^n)$$

The solution methods are presented in geometric fashion, i.e. as a completion of a square (Fig. 5.10). As the solutions were conceived as geometric entities, negative numbers were excluded: the sides of a square can after all never be negative. Numerous examples of such problems are given, sometimes involving polynomial fractions and/or irrational numbers (Fig. 5.9). In some instances, Mennher (with reference to Stifel) introduces new unknowns using the letters *A, B, C*.

5.5 Calculation of Interest

The solutions proposed in the arithmetic manuals generally show great faith in the underlying mathematical methods. But things were not all as they seemed, for, when it came to interest calculation, the authors were treading on uncertain ground. Of all solution types, these were arguably the most controversial.

On a broad sheet, seemingly intended to be displayed at an office, Maarten vanden Dycke explains how to calculate interest.[39] One of the problems is:

> "A pound of nails costs me 9 st.; at what price do I need to sell to make a 10% profit?"

[39] vanden Dijcke (1591/1592).

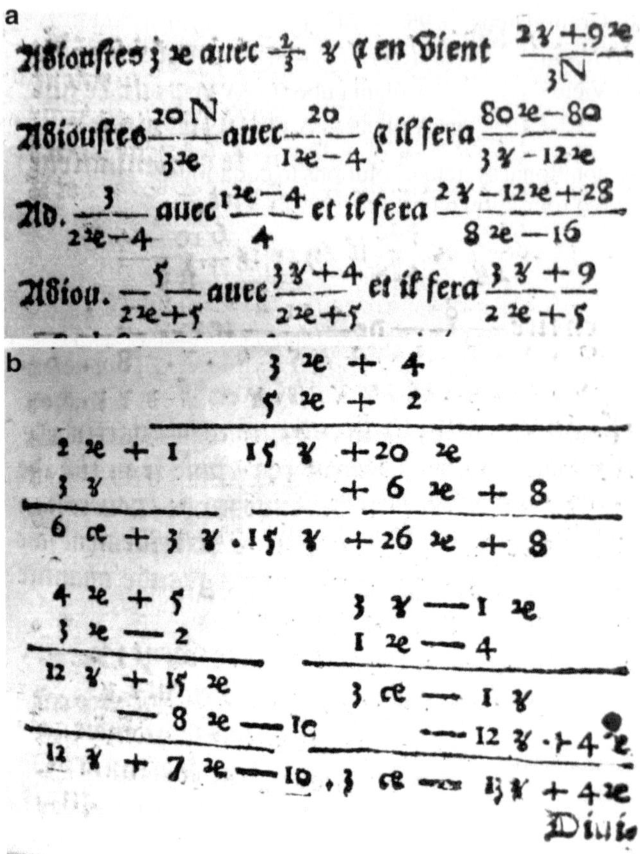

Fig. 5.9 From Mennher (1556) (EHC G 21409). (**a**) Multiplication of polynomials in cossic notation. (**b**) Calculations with polynomial fractions in cossic notation

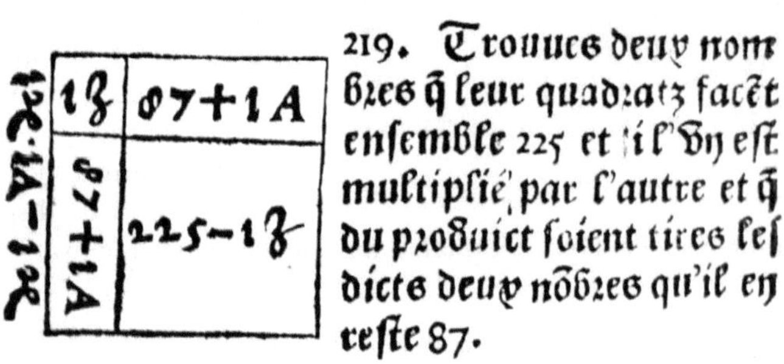

Fig. 5.10 Problem and explanatory figure leading to a quadratic equation (Mennher 1556, EHC G 21409)

The answer to us is $9\frac{9}{10}$, but according to Vanden Dycke it is 10 st. His reasoning goes as follows: "If I buy 100 st. worth of nails at 9 st./lb, I obtain $11\frac{1}{9}$lb; if I subsequently sell for 100 st. at 10 st./lb, I have sold 10 lb. But I also retain $1\frac{1}{9}$ lb, so that my profit amounts to $1\frac{1}{9}$ lb × 9st. = 10 st., and 10 st. is 10 % of 100 st.

Strictly speaking, his reasoning is not incorrect; only, he calculates the required 10 % profit on the basis of the selling price rather than the buying price. Unusual though it may seem now, it was common practice at the time (including in financial transactions).[40] Of course, all this is a matter of definition: what exactly does "profit" mean? Is profit the difference between the selling and the buying price? Or is profit the amount of goods one retains for one self after the resale has been concluded? Expressing profit in terms of goods is less abstract than expressing it in terms of money. Moreover, there was apparently no clear conception yet of what profit meant in a monetary economy. This is confirmed by Michiel Coignet's comments on the method in his amended version of Mennher's book (1573), where he actually stresses the difference between profit in goods and profit in money.[41] It is also an indication that the transition from a barter economy to a monetary economy was ongoing.

This was known to Mennher as well:

> Vn Marchant donne L.400 à interes pour vn an à 8. pour cent par an, pour cela il reçoit vne lettre de .432. L. & incontinent il est constrainct de reuendre la lettre, de cela il luy fault perdre .8. pour cent sur la somme de la lettre, combien d'argent recepura [sic] il content? Dictes, pour 100 de la lettre il reçoit L.92 content, combien recepura [sic] il pour .432? facit L. 397. β.8. ϑ.9$\frac{3}{5}$. Mais s'il perdoit .8. pour cent sur l'argent côtent, on le feroit par ceste Regle, en disant .108. font .100. combien font 432? Facit L.400 le premier capital. Ceste Regle doibt estre bien entendu, & m'ha estée souentefois demandée par les Marchants, à cause que l'vn veult compter l'interes côme l'argent centent vault sur la Bourse, & l'autre le veult compter sur la somme de la lettre, pource ceste discorde ne procede point de la calculation, mais par la faulte que l'vn n'entend l'autre.
>
> *A merchant lends £400 at an interest of 8% per annum, for which he receives a bill obligatory of £432. He becomes insolvent and is obliged to sell the bill. By selling he loses 8% on its value. How much money will he recuperate?*
>
> *Say: for £100 of the bill, he receives £92. How much will he recuperate for £432? Answer* facit L. 397. β.8. ϑ.9$\frac{3}{5}$.
>
> *But if he loses 8% on his money, applying this Rule would read as follows: 108 yields 100; then how much will £432 yield? Answer £400, the initial capital.*
>
> *One should have a good understanding of this Rule. Merchants have often asked me to explain it to them. The reason being that some merchants base the calculation on the money, as is customary at the Bourse, while others calculate it on the basis of the value of the bill. Their disagreement therefore does not stem from the calculation, but from the fact that they do not understand each other['s methods]*
>
> (Mennher 1565, Gviij*r*).

[40] De Groote (1967a, p. 206).

[41] Vanden Dycke's method is based on the books by Godevaert Gompaerst, whose work he edited during the 1590s. It also shows that the quality of books published after the Sack of Antwerp was low: old books with outdated methods were reprinted.

Another difficult problem which was still unsolved was the calculation of a nominal interest, despite Stevin's book.

E.g.

Een Coopman gheeft op interest lb 200 voor een iaer te 10 ten hondert opt iaer tellende alle 3 maenden interest op interest. Die vrage is hoe veel den interest beloopt? Antwoort: Wintmen in 12 maenden 10 soo salmen in 3 maenden winnen $2\frac{1}{2}$. Die welcke addeert tot 100 ende daer sullē comen $102\frac{1}{2}$. Die reduceert tegen 100 ende sullen comē 200 tegen 205 gelijc 100 tegē $102\frac{1}{2}$.

A merchant lends £200 for one year at an interest of 10% per annum. The interest is compounded quarterly. Find the amount of the interest earned.

Answer: If the interest on 12 months is 10%, then the interest on 3 months will be $2\frac{1}{2}$. *Add* $2\frac{1}{2}$ *to 100 and you find* $102\frac{1}{2}$. *Divide 100 by this number and you find that 200 is to 205 as 100 is to* $102\frac{1}{2}$,

	200
	205
200	41000
200	205
40000	8405000
200	205
8000000	1723025000
200	205
1600000000	353220125000

(Raets and Coignet 1580a, f G v *r*).

The calculation of compound interest is correct, given a nominal interest of 2.5 % per quarter. In fact, though, the nominal interest on 3 months with an effective interest rate of 10 % per annum would amount to around 2.4 % per quarter, not 2.5 %. The calculation is however in agreement with prevailing practice, as it was customary to put the nominal interest of a loan at an effective interest rate of x % per annum, repayable in instalments of $\frac{1}{n}$ years equal to $\frac{x}{n}$ %.

In the correspondence between Michiel Coignet and Bartholomeus Cloot, we find similar examples of confusion regarding interest calculation.[42] In one instance, Cloot asks Coignet how much one must repay if one lends £5,200 at 14 % per annum and redeems £500 per year. According to Coignet, at the *n*th installment, this amounts to $£\left(500 + \frac{500.n.14}{100}\right) = £(500 + 5n.14)$, with the exception of the last payment. Coignet thus repays £500 and calculates the interest due on this repaid amount. In his response, Cloot argues that words were forgotten in the problem formulation and that he meant "interest upon interest", as "was customary" in Holland. Coignet's method amounts to repayment with amortization with constant capital, in which the interest is calculated on the capital repaid at the *n*th installment. However, Cloot's method cannot be repayment in constant annuities, with an

[42]BNP Ms Néer 56, f18*v*–21*r*.

annuity equal to £500. For the first year, the interest on the outstanding capital amounts to £728, which is more than the proposed repayment. Hence Cloot can only mean constant capital repayments with an added interest on the outstanding capital. In that case, the two methods are equivalent.[43] Repayment in constant annuities was not uncommon in Antwerp: both Erasmus Schetz and Jan-Karel Affaitadi used the technique for loans to the City Council.[44] Nonetheless, the mathematics involved did apparently pose a problem for arithmetic teachers.

It was only in 1582 that Plantin first published Stevin's *Tables of Interest*. Such tables had existed previously as manuscript copies, but these "were guarded as big secrets by those who [had] them, and who [kept] them hidden".[45]

Problems which Stevin treats are:

> Determine the present value (or principal) A_n which is saved at $i\%$ for n years and has increased to a value of 10^7, i.e. determine A_n such that $A_n(1+i)^n = 10^7$.

Notice that the use of fractions is minimized by using 10^7. If the future value equals FC it suffices to multiply the result A_n with $\frac{FC}{10^7}$. These values are tabulated in a first series of tables. In a second series of tables the sum a_n of the amounts in the first column is given. This results in the invested amount on n years if one wants every down payment to reach 10^7. These tables are given for $i = 1, \ldots, 16$ and $n = 1, \ldots, 30$ and tables for penny 15 to 22.

Stevin only gives that values of A_n and a_n , i.e. the present values for compound interest of an annuity, which Stevin calls unprofitable interest. Tables for profitable interest, or future values S_n and s_n are not included.

These can of course be easily calculated using

$$\frac{1}{a_n} = \frac{1}{s_n} + i$$

$$\frac{1}{A_n} = S_n$$

One may wonder whether Stevin's reserve may have been induced by the episcopal prohibition on compound interest.[46] Of all the Antwerp arithmetic authors, only

[43]With Coignet's method, we have : $FC = C_n \sum_{j=1}^{n} \left(1 + \frac{i.j}{100}\right) = C_n \left(n + \frac{i.n(n+1)}{200}\right)$, while Cloot's method yields: $FC = C_n \sum_{j=1}^{n} \left(1 + \frac{i.(n-j)}{100}\right) = C_n \left(n + \frac{i.n(n+1)}{200}\right)$.

If the effective interest was 6.25 % (= penny 16), the nominal interest on 6 months would be put equal to 3.125 %. For an effective interest of 6.25 % per annum, the nominal interest per semester would be 3.078 %.

[44]De Groote (1968c, pp. 205–206). SAA T1242, f7*r*.

[45]Stevin (1582), fA4*r*. Jean Trenchant did publish a couple of tables in his *Arithmétique* (Lyons, 1558), but Stevin was the first to provide a complete set. Dijksterhuis (1970, p. 20), Zemon-Davis (1960, p. 19).

[46]See footnote 7 on page 10.

the Protestant Mellema makes reference to the immorality of compound interest, sounding the following warning:

Amy & Chrestien, tel probleme faisant
De ton frere gaigner ne fault penser autant
Car nous le proposent, non pour imiter
Tels meschans vsurier, mais pour les euiter.
Friend and Christian, such a problem might make
you gain from your brother, but do not think of it
for we propose it here not in order to imitate
the vile usurers, but to avoid them

Mellema insists that he merely gives the examples to draw the readers' attention to this practice, so that they could refrain from it. Peter van Halle, in his manuscript,[47] similarly points out that, while compound interest is usury and unfit for Christians, it does exist, and hence he feels compelled to instruct the reader in its techniques.

In the 1573 edition of Mennher's *Arithmétique,* we find the following problem:

Un marchand donne £ 100 a interest a 16 pour cent par an, et au bout de chasque 3 mois veult il auoir le 1/4 du principal et interest, et que les payementz soient egaux: on demande, combien chacun payement monte. $27\frac{1}{2}$.

A merchant lends £100 at an interest of 16% per annum. Repayments are due quarterly. Each repayment consists of capital and interest. All repayments must be equal. Which sum is to be repaid quarterly?

(Mennher and Coignet (edit) 1573, f k *r*)

Mennher put the interest per quarter at 4 %. Thus he arrives at four interest calculations: $\frac{104}{100} = \frac{1}{25/26}$, $\frac{108}{100} = \frac{1}{25/27}$, $\frac{112}{100} = \frac{1}{25/28}$ and $\frac{116}{100} = \frac{1}{25/29}$. The sum of these four fractions yields $3\frac{366178}{570024}$ and he finds $3\frac{366178}{570024} : 1 = 100 : £\,27–9–1\frac{659}{8505}$ for each payment.[48] Coignet disagrees: he puts the solution at £27-10. The actual solution to the problem is 27.549005 or very close to £27-11.

Problems involving interest were such that they occupied the minds of many mathematicians for a long time.

Vn Seigneur a vne rente de £1000 a payer en 5 ans, lesquelz il veut receuoir en 5 payementz, a scauoir, chascun an vne somme, & que l'vne soit egale a l'autre: on demande combien chascun payement monte, quand on conte $3\frac{1}{3}$ pour cent par an, a conter tous les ans interest sur interest.

A gentleman has a bill obligatory of £1000. This sum of £1000 has to be repaid over a period of five years. The gentleman wishes to receive five equal payments. The interest is $3\frac{1}{3}$*% compounded [annually].*

Which sum is repayable annually?

(Mennher 1565, f k iij *r*)

According to Mennher, the solution is £187-2-$5\frac{60945}{109715}$.

[47] van Halle (s.d.), f205*r*.

[48] De Groote (1968c, pp. 203–204).

In Coignet's revised edition (1573), he asserts that Mennher made a calculating error, and gives as the (correct) solution £187-2-0$\frac{3803496}{4329151}$.[49] Coignet used coss to solve the problem, i.e. he solved the equation $x\sum_{j=1}^{n}(1+i)^{n-j} = 1,000$.

Marten Wentsel, a North Netherlandish arithmetic teacher in consecutively Rotterdam, Amsterdam and Middelburg, copied the problem in his *Proportionale Ghesolveerde Taffelen van Intrest* (1594).[50] Unfortunately his solution is erroneous.

Wentsel had encountered the problem in Mennher's book, but the direct inducement for its inclusion was a question that had previously been put to him by Felix van Sambeke (a teacher who had emigrated from Antwerp). Van Sambeke had sought Wentsel's help in 1586, a year after leaving Antwerp, saying that he was able to apply the technique, but failed to understand it. Wentsel had answered the question by giving a solution, which Van Sambeke (rightly) identified as erroneous. Wentsel was deeply insulted, and therefore published the problem and its solution in his book and challenged Van Sambeke to disprove his solution.

According to Wentsel each installment had to equal

$$\frac{1}{5}\left(\sum_{0}^{4}\frac{\frac{1,000}{5}}{(1+i)^{j}}\right) \qquad \left(\text{in general: } \frac{1}{n}\sum_{0}^{n-1}\frac{\frac{S}{n}}{(1+i)^{j}}\right)$$

What he did was to calculate the equivalent of 200 gl., i.e. the amount that, if invested at an interest of $i\,\%$, would yield 200 gl. at the end of the term. The sum of these five amounts he divided by five to calculate the constant annuity. The method fails because it does not calculate the interest on the actual invested amount, but on the amount that would be invested in the case of a non-constant annuity with equal final capital. Ludolf van Ceulen accepted the challenge and, in his book *Vanden Circkel* (1596), arrives at the same solution as Coignet. Wentsel stubbornly stood his ground, complaining in his book entitled *'t Fondament van Arithmetica* (1599) that his reputation had been tarnished, and at once launching a scathing attack on Van Sambeke "who had brought [him] this venom".[51] Wentsel complains repeatedly about Van Sambeke's ignorance, and also implicated another Antwerp schoolmaster, Anthony Smyters, who had apparently visited him in 1595 to warn him he should stop treating Van Sambeke in this way. Wentsel did not heed the call, however, and argued that friends were shielding each other at the expense of the truth. He never conceded ...

---

[49]Mennher (1565), Mennher and Coignet (edit) (1573), k iij *r*. The yearly instalment is $1,000.\frac{1}{(1+i)^5-1} = 187{,}10368\ldots$ Conversion into pounds yields £187 2 st. 0$\frac{882}{1,000}$, which is equal to Coignet's solution, save a rounding error.

[50]Waller-Zeper (1937, p. 66ff).

[51]UB Amsterdam 967C16, f115 and 158.

5.6 Higher Order Equations

In his 1565 *Pratique* Mennher solves a couple of third degree equations. Some of these lead to the relatively easy to solve type $x^3 = a$. Only one leads to a Cardan-Tartaglia type equation $38.5x - x^3 = 90$.

> Item il y ha vn corpz columnaire d'vne egale grandeur, duquel le diametre de par dedans fait 7, & la hauteur 5, & il peut en tout côtenir 500 potz de liqueur, mais il n'est point tout plain, c'est à dire, qu'il ne côtient q̄ $233\frac{59}{77}$ potz, & puis il y ha vn cubus de telle grandeur quand on le met dedans la liqueur, la hauteur de la liqueur fait iustement autant qu'vn costé dudit cube. la demande est, combien soit chascun costé dudit cube? Responce: Posez la hauteur du cube soit 1ꝛe, sa grandeur sera 1ce & si on multiplie $38\frac{1}{2}$ l'aire de la circunference, auec 1ꝛe, il en vient $38\frac{1}{2}$ꝛe, desmesmes tirez 1ce, & fera $38\frac{1}{2}$ꝛe - 1ce Plus, multipliez $38\frac{1}{2}$ l'aire, par 5 la hauteur, & en viendront $192\frac{1}{2}$ pour lan grâdeur de tout le corpz columnaire. Dites, 500 potz ont $192\frac{1}{2}$ de grandeur, combien ont $233\frac{59}{77}$ de grandeur? facit 90, qui sont egaux à $38\frac{1}{2}$ꝛe - 1ce & 1ꝛeest egal à 4 pour chascun costé dudit cube (Mennher 1565, f Pp vij *v*-viij *r*).
>
> *Item a columnar body has an inner diameter of 7 and a height of 5. This body can contain 500 pots of liquor. It is not filled completely, but contains just* $233\frac{59}{77}$ *pots. A cube is put [horizontally] into the liquor. The height of the liquor rises to the height of the cube. What is the size of the side of the cube?*
> *Put the side of the cube equal to* x *then* $V = x^3$
> *and* $38\frac{1}{2}x - x^3$ *=volume of liquid*
> *192.5 unit*^{3}*holds 500 pot*
> *therefore 90 unit*^{3} *holds* $233\frac{59}{77}$ *pot*
> *Whence* $38\frac{1}{2}x - x^3 = 90$
> *with* $x = 4$ *as a solution* $\left(\textit{and other solutions } x = \dfrac{-8 \pm \sqrt{424}}{4}\right)$

It is the only problem in the book leading to a third degree equation. No example of a solution method is provided. Whereas for quadratic equations, the algorithm is clearly explained, no such explanation is given for a third degree equation. In this instance, the reader is expected simply to accept Mennher's reasoning.

In 1577, Peter Heyns presented Coignet with a question that had been sent to Wouter de Coster by an unknown correspondent in Cologne.[52] Apparently neither Heyns nor De Coster were able to solve the problem. The question goes as follows:

> Deylt 8 in twee deelen dat alsmen dmeeste ghetal multipliceert met tquadraet vant minste datter wt come 72.
>
> *Divide 8 into two numbers such that if the largest number is multiplied with the square of the smallest number the product equals 72.*

[52] Coignet (1576–77), f12*r*-*v*.

Coignet noted that the problem was insolvable in its proposed form. He argued that if the solution were 4, being the half of 8, the desired product would be $4.4^2 = 64$, which would be a maximum.[53]

Therefore, the product could not be 72. Coignet subsequently argued that the question should read:

> *Divide 8 into two numbers such that the square of the largest number multiplied by the smallest equals 72.*

This leads to the equation

$$x^2 (8 - x) = 72$$

or $x^3 + 72 = 8x^2$ (1ce + 72 Egal 8ʒ)

which has $x = 6$ as a solution.

> *To easily find the solution to this equation, bear in mind that 1ʒe cannot be 8, but has to be 8 or 6 or 4 (that is the factors of 72). Insert these numbers into the equation one by one. You will find that the sum of 216 [= 6^3] and 72 is 288, while 8 multiplied by 36 [= 6^2] is also 288.*

In this case, the solution is found by "guessing", which is possible so long as the solution is an integer number. To conclude from this that Coignet did not know the algorithm for solving third degree equations is however premature. When dealing with altogether more complex partnership problems (see p. 86), the equations are less likely to be fortuitously solved.

5.7 Partnership Problems

Many of the problems in the arithmetic books concern the emergence of a company, a partnership of men banded together in an ongoing money-making exercise. They would each put forward a particular amount towards the sum required to establish the company in the hope of subsequently earning a return on their investment. In Antwerp, this would typically concern a share in a cargo company or an enterprise specializing in overseas trade with, for example, Russia or the Baltic.

The partnerships in question were very often limited in time, with different merchants contributing different amounts for varying periods of time. This gave gave rise to a popular kind of problem known as "partnership problems". These typically involve two or more merchants who invest in a venture and subsequently withdraw their money at different moments in time. The question that arises is how should the profit, once it is known, be divided between the various partners.

---

[53] I.e. for $y = x(8 - x)^2$ and $0 < x < 8$, $y = 64$ is a maximum.

A classic example is:

Deux marchans font compa. l'Un mect £300 après la reprinse il adiouste £100 et auec la reste il demoure iusques à la fin de l'an. Le second mect £600 & apres 2 mois il reprend £200 & 2 mois apres la reprinse il reprend encore £200 et auec la reste il demoure l'an entier et ilz gaignet ensemble £300/co(m)bien de gaing prendra chacun.

2..300.. 600
100
2..200.. 400
100
8..300..2400 3400
2..600..1200
200
2..400.. 800
200
8..200..1600 3600
7000..300..3400 faict £145$\frac{5}{7}$ po^r le premier
7000..300..3600 faict £154$\frac{2}{7}$ po^r le second

Two merchants form a company. The first initially invests £300, the second invests £600. After two months the first withdraws £100, while the second withdraws £200. After another two months the first reinvests £100, while the second again withdraws £200.

Their investments now remain unchanged until the end of the year. At the end of that year they have gained £300 on their investment. How should the proceeds be divided?
[...]
7000..300..3400 gives £145$\frac{5}{7}$ for the first
7000..300..3600 gives £154$\frac{2}{7}$ for the second
(*Mennher 1556*, f Ei *r*)

Mennher first determines the sum of the amounts invested per month by each of the two merchants. These sums determine the relative share in the profit that each merchant is entitled to. In its simplest form, one has to find the proportional shares, so that the solution is given by:

$P_i = P\dfrac{C_i}{\sum C_i}$ where P is the profit, P_i the profit for the i-th merchant, and C_i the invested capital by the i-th merchant.

If the merchants invest their money for different periods of time, the profit becomes:

$$P_i = P.\frac{C_i t_i}{\sum C_i t_i}.$$

If a merchant invests different amounts of money for different periods, this becomes:

$$P_{i,j} = P.\frac{C_{i,j} t_{i,j}}{\sum C_{i,j} t_{i,j}} \text{ and } P_i = \sum_j P_{i,j}.$$

Coignet (1576–77), f 33*v*, solves the problem using equations:

Item twee legghen tzaemen in compaignie £12 en handelen daermede eenen sekeren tyt te A 3 maenden en B 6 maenden ten eynde des tyts vinden zy £27 in capitael en winnighe hieraff neempt A £7 en B £20/voor Capitael en winninghe: vraeghe hoeveel elck ingheleyt heeft? Facs. A 4£, B 8£.
Two merchants [A and B] form a company with a combined investment of £12, which is invested. A stays invested for 3 months, B for 6 months. By the end of this period, the capital of the company has grown to £27, consisting of the initial capital and the profit. When this sum is divided, A receives £7, while B receives £20. What was their respective initial investment in the company? Ans. A 4£, B 8£.

By putting the amount that A invests equal to 1℞ (= x), B's investment equals 12 – 1℞ .

By multiplying these amounts with the periods of investment, one can find the relative share of each.

A invested his money for 3 months, making $3x$,
B invested for 6 months, making $6(12 - x) = 72 - 6x$.
The total profit P is 15, A's profit is $P_A = 7 - x$.
The total investment was $\sum C_i t_i = C_B t_B + C_A t_A = 72 - 6x + 3x = 72 - 3x$
The ratio profit/investment has to be equal for both, yielding the equation:

$$\frac{P}{\sum C_i t_i} = \frac{P_A}{C_A t_A} \Leftrightarrow \frac{15}{72 - 3x} = \frac{7 - x}{3x}$$

whence $(72 - 3x)(7 - x) = 3x.15 \Leftrightarrow 3x^2 - 138x + 504 = 0$
which is a quadratic equation with solutions $x = 42$ and $x = 4$. The first solution is obviously unrealistic, since it is larger than the company's final capital.

The equation is of course arrived at using the rule of three.

Such problems occasion complicated solutions, for which the resulting equation is not always a quadratic equation.

For instance, Coignet cites from a book by Simon Jacobs[54]:

Dry hebben tzaemen ingheleyt 240 *f* staen daer mede A 10 maenden B 7 maenden & C 4 maenden/naer eynde des tyts vinden sy in als 483 *f* hier aff neempt A 175 *f* B 164 *f* C 144 *f* in principael & winninghe. Vraeghe hoeveel elck ingheleyt heeft? Facs A 70 B 80 C 91.

Three merchants [A, B and C] form a company worth an initial capital of 240 guilders. A stays invested for 10 months, B for 7 months and C for 4 months. At the end of this period the capital has grown to 483 guilders. When this sum is divided among the merchants, A receives 175 guilders, *B 164* guilders and *C 144* guilders. *The question is how much did each one invest into the company initially? Ans. A 70 B 80 C 91* (Coignet 1576–77, f 30*v*, added note).

[54] Probably Simon Jacob(s) (1510–1564), a well known Frankfurt *Rechenmeister,* who authored *Ein New vnd Wolgegründt Rechenbuch*, Frankfurt a.M. (1565).

Coignet initially tries to solve the equation by using two unknowns. Putting A's capital at x and B's capital at y, C's capital becomes $240 - x - y$, which leads to the equation:

$$\frac{233{,}280 + 1{,}458x + 729y}{960 + 6x + 3y} = 243$$

It is obvious that this equation has an infinite number of solutions. Therefore, Coignet puts the annual profit for 100 st. (= 5 gl.) at x, which gives rise to a third degree equation:

$$3{,}393\frac{3}{4}x + 30{,}375 = 38\frac{8}{9}x^3 + 402\frac{1}{12}x^2$$

$$\Leftrightarrow x = 9 \text{ or } x = \frac{-1{,}083 \pm \sqrt{9{,}361}}{112}$$

Coignet merely mentions the first solution; he does not give a solution technique. Some light may be shed on this technique by another equation Coignet solves. This latter equation may be the solution to a problem, but this is not stated.

Item 1ce + 10ʒ + 20ꝛe + 20 Egal 9ʒ + 24ꝛe +24
treckt al elcke sijde aff 4ʒ + 8ꝛe + 12 soo
compter 1ce + 6ʒ + 12ꝛe + 8 Egal 5ꝛe + 16ꝛe +12
Deylt elcke partije met 1ꝛe+2 soo compt
1ʒ + 4ꝛe + 4 Egal 5ꝛe + 6
4ꝛe + 4 4ꝛe + 4
1ʒ Egal 1ꝛe + 2

Item $1x^3 + 10x^2 + 20x + 20 = 9x^2 + 24x + 24$
subtract $4x^2 + 8x + 12$ *from both sides and one finds*
$1x^3 + 6x^2 + 12x + 8 = 5x^2 + 16x + 12$

Now divide both sides by $x + 2$ *and one finds*
$1x^2 + 4x + 4 = 5x + 6$
$\Rightarrow 1x^2 = 1x + 2$
$\Rightarrow x = 2$
(Coignet 1576–77, f 46*r*, added note).

The latter equation is a quadratic equation for which the solution algorithm is of course well-known. In this case, we see that Coignet uses a technique equivalent to our factoring using Horner's rule. While in the previous equation the factor $x - 9$ (or the solution $x = 9$) could still be "guessed", this is not the case for $x + 2$. Using the remainder theorem (or any other rule equivalent to it) would imply counting with negative numbers. The question then remains whether Coignet and his contemporaries knew an algorithm that is equivalent to dividing polynomials.

5.8 Geometry

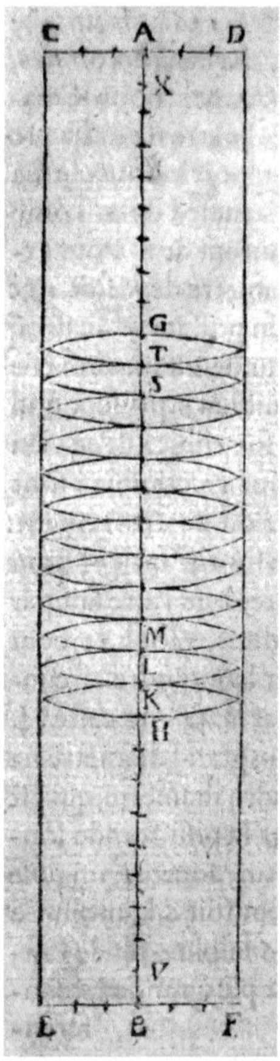

Fig. 5.11 Lunes (Mennher, 1565, MPM A3589)

Although the fourth part of Mennher's *Practicque*, dealing with geometry, begins with Euclidean definitions of points, angles, triangles and the like, from the very first problem it also involves the calculation of unknown quantities. The proposed problems are classical queries relating to the trigonometry of right-angled triangles involving fallen trees, ladders, etc, but also the construction of a pentagon, decagon and hexagon. Some give rise to equations expressed in cossic notations.

In some instances, Mennher goes further and uses sines. In one unexpected and particularly interesting problem he constructs lunes with which to cover a sphere (Fig. 5.11). To this end, divide the circumference of the sphere into 12 equal parts, A. Draw a line consisting of 30 of these parts A, and call the points A_i. In the points A_{10} to A_{22}. Construct a rectangle by extending perpendiculars $3A$ to each side. With a compass opening of $10A$ and one end in A_0, draw an arc through A_{10}. Now repeat the process by putting one end in A_1. Once 12 arcs have been drawn, repeat the procedure starting at A_{30}. Thus 12 lunes are found, which almost exactly cover the sphere.

Obviously the problem is inspired by the construction of globes, onto which such lunes would be glued.

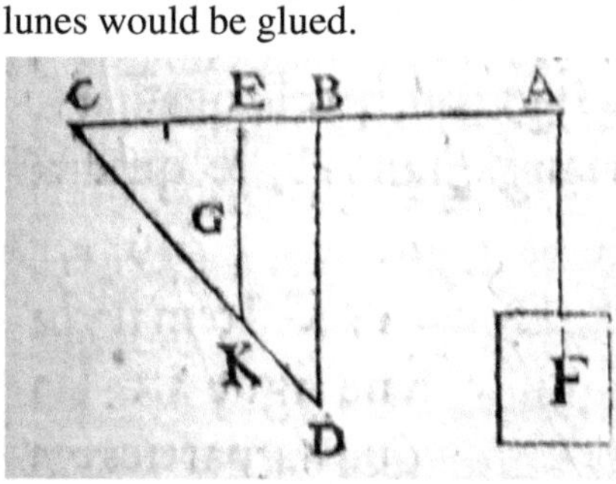

Fig. 5.12 Lever problem (Mennher, 1565, MPM A3589)

Other problems involve levers and centres of mass (Fig. 5.12). Given a lever AC with the fulcrum (pivot) in the midpoint B, if a square weight is suspended in A, and a right-angled triangle is suspended on BC, then the triangle and the square will have a ration of 3 to 1. Mennher correctly goes on to point out that if the triangle is suspended in the centre of mass, it can be rotated about this point at will without breaking the equilibrium of the lever.

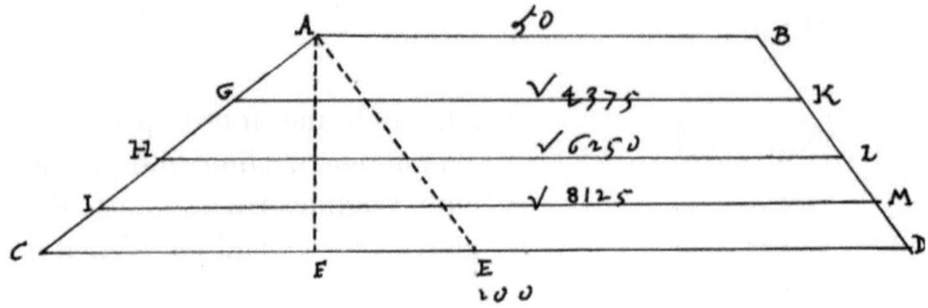

Fig. 5.13 After Coignet (1576–77)

Irrational solutions are not uncommon in this kind of geometric problem. In fact, they are quite common in problems starting out with realistic conditions.

E.g. Pauwelsen-Coignet 1577 (Fig. 5.13):

If in a trapezoid $\|AB\| = 50, \|CD\| = 100, \|AC\| = 40$ *and* $\|BD\| = 30$. *Three line segments are drawn in this trapezoid parallel with* AB, *in such a way that the areas of the resulting trapezoids are equal. Determine the area of the trapezoid and of the length of the line segments.*

The solution is: area AGKB = *450,* $\|GK\| = \sqrt{4{,}375}$, $\|HL\| = \sqrt{6{,}250}$, $\|IM\| = \sqrt{8{,}125}$.

5.9 Trigonometry

Valentin Mennher and Michiel Coignet were the only authors to write on trigonometry. Trigonometry-like problems are present in other arithmetic books, but these can invariably be reduced using Pythagoras' Theorem or properties of similar triangles (Fig. 5.14). They can, in other words, be reduced to equations with rational coefficients.

Trigonometry as we know it has its origins in Germany in the fifteenth century.[55] Georg Peurbach (1423–1461) and his student Johannes Müller of

[55] Practical trigonometry has been around since Ancient Times. The Greeks and the Arabs founded it in a comprehensive theory. The Germans were however the first to systematically draw up tables with the values of trigonometric functions. See Bond (1921), Maor (1998) and Kaunzner (2005b).

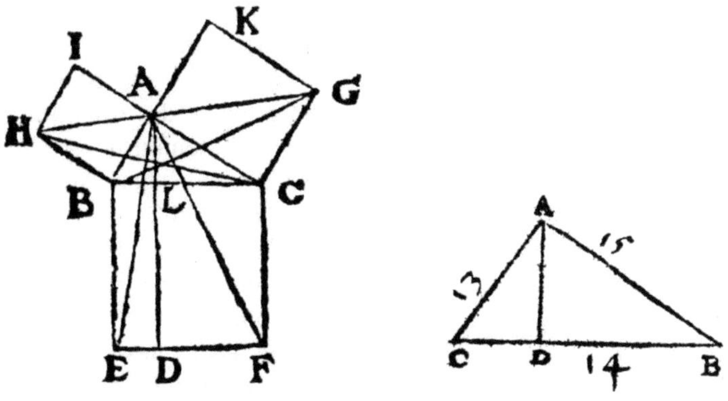

Fig. 5.14 Illustration for Pythagoras' Theorem and its application (Mennher 1556, EHC G 21409)

Koeningsbergen (Bavaria), better known as Regiomontanus, laid the foundations of modern trigonometry. Peurbach introduced the so-called Hindu-sine, i.e. he equalled the sine to the half chord of an angle.[56] Written in 1462–1463, *De Triangulis* (which was only published after his death in 1533) brings together all (plane and spherical) trigonometric knowledge in a coherent way. It contains the rule of sines for spherical triangles:

$$\frac{\sin a}{\sin A} = \frac{\sin b}{\sin B} = \frac{\sin c}{\sin C}.$$

The first book on trigonometry to be published in Antwerp was Frisius's *Libellus de locorum* (1533), a treatise on triangulation in surveying (see Chap. 9, p. 162). Other authors on trigonometry were -again- Mennher and Coignet, neither of whom were original in their printed work, drawing extensively as they did from the writings of Southern German authors. Most problems are practical, relating to the determination of longitude. Mennher's *Practique des Triangles Sphériques* (1564) begins with definitions, properties and some solved problems. The second chapter is devoted to spherical triangles and has a similar structure. There are no proofs in the book, only problems relating to the angular distance between stars, polar and solar

[56]Kline (1972, p. 328).

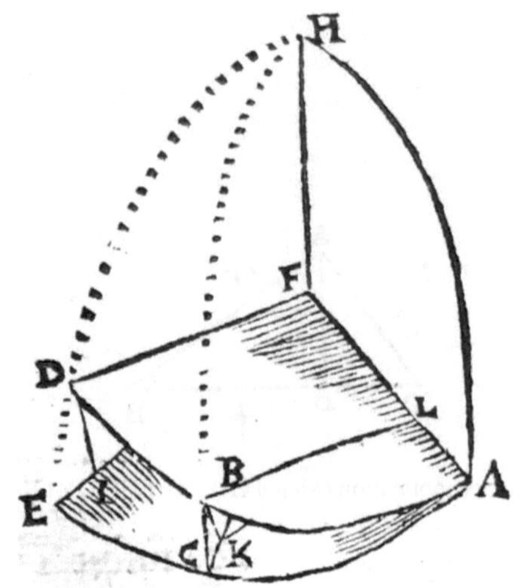

Fig. 5.15 Coignet (1573), MPM A 4348

altitudes and time pieces. Coignet (1573) refers to Regiomontanus, Copernicus[57] and Mennher. Because "le principe des triangles sphericques n'y est demonstre" (Civ *v*- v *r*), he proves that $\sin C/\sin A = \sin AB/\sin BC$ in a right-angled spherical triangle. His proof is different from that proposed by Regiomontanus'.[58]

He also cites the sine rule in a general triangle, but offers no proof, referring merely to Geber (Abū Mohammed Jābir ibn Aflah[59]).

---

[57]N. Copernicus, *De latribus et angulis tien planorum et rectilinearum tum sphericorum*, Wittenberg, 1542.

[58]For Regiomontanus's proof, see Smith (1959). Coignet's proof reads (See Fig. 5.15):

Let H be the pole of the circle AC
(i.e. we suppose that AC is the equator of the sphere).
Draw a great circle HE and AE ($AE = \pi/4$)

$$AB \cap HE = D$$

Let F be the centre of the sphere. Draw $BL // DF$.
Construct $BK \perp FC$ and $DI \perp FE$.
Because $FHBC \perp AFE$ and $BK \perp AFE$
and $FHDE \perp AFE$ and $DI \perp AFE$
we have $DI // BK$.
Whence $\Delta FDI \cong \Delta LBK$,
because $\measuredangle BLK = \measuredangle DFI$ en $\measuredangle BKL = \measuredangle DIF = \frac{\pi}{2}$,
from which: $\|FD\|/\|DI\| = \|LB\|/\|BK\|$
Now

$$\|FD\| = \|HB\| = R = \sin C$$
$$\|DI\| = \sin A$$
$$\|BK\| = \sin BC$$
$$\|BC\| = \sin AB$$

whence $\sin C/\sin A = \sin AB/\sin BC$. qed

[59]Arab mathematician, active in Seville between 1140 and 1150. His work was translated into Latin by Gerard of Cremona (1114–1187).

5.10 Foreign Influences

It is not easy to trace external influences. For an arithmetic book, one can proceed on the basis of cited books or by comparing the problems posed to problems in other works. Christoffel Rudolf is the most cited author in arithmetic books, but by a small number of authors: Frisius (1540), Mennher (1565) and Coignet (1587). Mennher demonstrates his familiarity with other books when he relates the well-known tale of Archimedes and the supposedly golden crown. He states that Cardano solves the problem using the Rule of Algebra and in the cossic notation, whereas Frisius solves it by relying on false position, and Simon Jacob uses the "Partnership Rule" (i.e. implicitly solving a system of equations).

Coignet mentions a few other names, all classical and Netherlandish authors, with the exception of Simon Jacob, Christoffel Rudolf and Giorgio Valla. Only the wine gauging treatise by Raets-Coignet provides a clear clue that Italian authors such as Luca Pacioli and Tartaglia were known in the Low Countries. However, their direct influence seems to have been small. Rudolf's and Stifel's influence is perhaps most apparent in Mennher's use of symbols for irrationals (See Fig. 5.16).

His notation for roots ($\surd$ and $\sqrt{\mathfrak{z}}$) was probably taken from Stifel's *Arithmetica Integra* (or a common or intermediary source[60]) where one finds $\sqrt{\mathfrak{z}}$, $\sqrt{\mathfrak{c}}$, $\sqrt{\mathfrak{zz}}$...[61] for $\surd$, $\sqrt[3]{}$, $\sqrt[4]{}$ respectively. Mennher does not use $\sqrt{\mathfrak{z}}$ consistently, but also writes $\surd$. The latter notation is found in the books by Christoffel Rudolf. Mennher's student Coignet used $\surd$ and sometimes $\sqrt{\square}$, while for the cube root he used $\sqrt{cb}$.[62]

Mellema's list (1587) of deceased mathematicians "de nostre temps et siècle" is a case in point: no fewer than 16 of the 29 names are indicative of a German

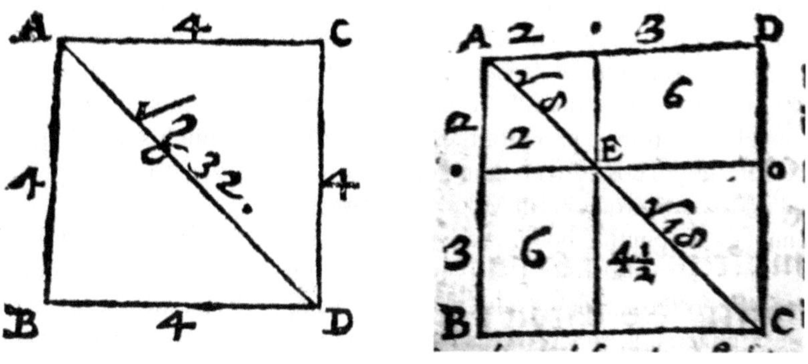

Fig. 5.16 Irrationals in Mennher (1565) (MPM A 3589)

---

[60]The sign $\surd$ was first used in Germany. It was applied in a manuscript by Adam Riese entitled *Der Coss* and seems to have evolved from a sign resembling a point. Cajori (1993, p. 366 ff.).

[61]Tropfke (1980, p. 296).

[62]See for example BNP Ms Néer 56.

descent (not counting Mennher). The conclusion is obvious: the Antwerp scene was influenced profoundly by Southern German sources, which is not surprising considering the close trade relations between the two regions.

Some arithmetic books drew a lot of interest and were plagiarized in other countries. Numerous editions of Frisius were printed in other European towns.[63] It goes without saying that subsequent editions of books originally produced in Antwerp frequently appeared in towns in the Republic, in consequence of the massive emigration wave. Many schoolmasters and printers who had fled Antwerp found a new and eager market here. Mennher's *Livre d'Arithmétique* (1561) was republished in the Republic by the emigrated printer Jan van Waesberghe in 1589 and 1609. Mennher also influenced Northern mathematicians via Petri's *Practique om te leeren cijferen* (1583; 1591; 1596; 1598; 1603; 1605).[64] The 1583 edition of this book was in fact a reworked version of the 1567 edition, extended with an algebraic section. Many problems in this book drew inspiration from Mennher's *Practique pour Brieuuement apprende à Ciffrer.* Petri proposed trigonometric problems he had encountered in *Practique des Triangles Sphériques* (1563), with latitudes and longitudes adapted to Amsterdam instead of Antwerp. He seems also to have drawn from Mennher's work in his writings on accounting. Mennher's books, together with the work of Raets-Coignet, also provided the inspiration for a number of manuscript manuals.[65]

Although the local population remained the primary market, some books would find their way to towns a thousand miles away. For instance, the library of the architect Hans Schiller in Leipzig contains a copy of "a niederlendisch rechebuch" (a Dutch arithmetic book).[66]

Valentin Mennher's books were published at least twice outside the Netherlands. Moreover, his 1560 book in German was intended for the German market. In 1555, E. Barricat published a *Instruction D'Arithmétique Povr Briefvement Chiffrer et Tenir Liures,* which seems to have been a full reprint of Mennher's *Practique Briefue pour Cyfrer et Tenir Liures de Compte,* published in 1550 at Jan van der Loe's. It was common practice to literally re-print books that had been published elsewhere and that held the promise of a profit. Christopher Plantin built his fledgling business on such reprints.[67] Possibly an Antwerp copy found its way to Lyons (which was situated on the trade route from Milan to Antwerp).

Mennher's accounting books were published in a Spanish translation by Antich Rocha[68] in 1565 in Barcelona. It is no wonder that a Netherlandish book should show up on the Iberian peninsula, given the strong dynastic ties between the two countries. In fact, it is rather surprising that so few volumes found their way there.

---

[63] See van Ortroy (1920) and Reich (2005).

[64] Bosmans (1908b) and Kool (1999, pp. 263–264).

[65] See Kool (1999, pp. 263ff.) for a detailed analysis of the network of books influenced by Mennher's books.

[66] Rüdiger (2005, p. 428).

[67] See Voet (1989).

[68] Born in Girona, Catalonia, he produced numerous translations and digests.

John Tapp's *The path-vvay to knovvledge* (1621) is partly based on Guillaume Gosselin's French translation of Niccolò Tartaglia and partly translated from Valentin Mennher.[69] There is still a resemblance between the problems in Thomas Simpson's *A Treatise of Algebra* (1745) and Mennher's problems.[70] Undoubtedly via John Tapp's books, these problems found their way into many English publications on arithmetic.[71] Other works on arithmetic from Antwerp found their way to England, e.g. François Flory's *Les Practiques de Chiffre* (1577), which was translated by John Weddington and published in 1593 as *The Practize of Ciphering*. It is a near-literal translation in which even the pre-1576 exchange rates are copied.

In 1591, Anton Neudorfer went to Cologne to study French at Adrian Dennsten's. To demonstrate his ability, he translated Mennher's 1570 *Arithmétique* into German, a rendering that unfortunately never seems to have been published.[72]

In France, as we have previously mentioned (see p. 59), Denis Henrion plagiarized Mennher and Coignet's books in the 1620s.

There appears to have been a lively correspondence among the Netherlandish arithmetic teachers. These contacts are very important to the history of mathematics, as the competition that was thus engendered sometimes paved the way for the formulation of new theories. Such correspondence could also lie at the heart of efforts to find a justification or proof of assertions. Episodes such as that featuring Marten Wentsel (see p. 81) make it clear that he needed convincing. This could only be achieved by stating a general method and proving its mathematical soundness.[73]

One source for this correspondence is a kind of mathematical diary kept by Michiel Coignet. In the collation, different pieces are brought together, including notes from the period 1577–1578. It details, for example, that Peter Heyns sought Coignet's advice on a problem, which the former had received from Wouter de Coster, in April 1577.[74] De Coster had in turn been presented with the problem by a correspondent from Cologne, whose name is not mentioned. Neither Heyns nor De Coster were able to solve the problem, or, more correctly, they had both failed to observe that the problem formulation was flawed. Coignet immediately spotted the error, corrected it and solved the problem by means of a third degree equation (see p. 83).

---

69 Huntington Library, Rare books, 69609.

70 Heeffer (2005, p. 6).

71 On mutual influences between textbooks, see for example Heeffer (2005, 2006b).

72 Haller (2008, p. 104). It is not clear whether Adrian Dennsten and Adriaan Dens were one and the same person. The latter was an Antwerp arithmetic teacher and who is last mentioned in Antwerp in 1580. Hendrik Dens Adriaansz. left Antwerp for Cologne in 1584. See De Groote (1967a, p. 249).

73 Note that we also refer to a justification. In general, the arithmetic teachers were satisfied with a general example. Others demanded a proof. Since there was no real axiomatic system, we cannot speak of a formal proof.

74 Coignet (1576–77), f16*r*.

Fig. 5.17 Coignet's introduction to Jan Pauwels's problem. Coignet's manuscript reveals that Pauwels was Ludolf van Ceulen's mathematics teacher

In April 1578, Coignet received a couple of problems from Bartholomeus Cloot.[75] Coignet sent him an answer and added two questions of his own. One of these he refers to as "the question from Cologne", which seems to indicate there was a continuing correspondence with the German town. In fact, it should not come as a surprise that contact was maintained between arithmetic teachers from Antwerp and Cologne, as not only were the two cities linked by a trade route, but during the reign of Alba many Antwerp arithmetic teachers had fled to the Rhineland.

It was through Cloot that Coignet came into contact with Ludolf van Ceulen, who sent Coignet the solution to the problem from Cologne.[76] The two must have kept in touch, as is apparent from a letter that Coignet wrote to Galileo in 1588, and in which he presents the latter with a problem formulated by Van Ceulen.[77]

Coignet also came into contact with Jan Pauwels of Middelburg, who was Van Ceulen's mathematics teacher.[78] According to Coignet, Pauwels was a *geometer*, i.e. a surveyor. The first problems Pauwels sent to Coignet all relate to surveying (Fig. 5.17). The two men seem to have maintained a lively correspondence.

All the problems presented, with the exception of the one from Cologne, originated from Netherlandish correspondents, and they are all typical for arithmetic teachers. In that sense, the book is representative of the kind of contacts that existed between arithmetic teachers.

Coignet also maintained contacts with other well known or foreign mathematicians, including Galileo.

Abraham Ortelius informed Coignet about Galileo's work on the centres of gravity of frusta of paraboloids. This prompted Coignet to write a letter to Galileo, making him Galileo's first foreign correspondent, in which he presents the latter with his own proof (based on Galileo's lemma).[79] Galileo's contribution to the analysis of centres of gravity is based on a theorem that says that if the weights in arithmetic progression are equally spaced along the arm of a balance, their centre of

[75]Coignet (1576–77), f18*v*.

[76]Coignet (1576–77), f21*v*.

[77]Galileï (1968, vol. 10, p. 31ff.).

[78]Coignet (1576–77), f22*r*.

[79]Galileï (1968, vol. 10, pp. 31–33).

gravity divides the balance arm in the ratio 2:1.[80] The argument obviously bears some relation to certain problems posed by the arithmetic teachers (see p. 87). Coignet's letter came very soon after Galileo had made his results known. He had worked on the centres of gravity of parabolic solids during 1587. Coignet was very lauding about Galileo's extension of Archimedes' work, especially in relation to its practical importance. Coignet further laments that no one in the Netherlands seemed to be interested in mathematics anymore.[81]

He also posed a problem he had received from Ludolf van Ceulen concerning the lengths of certain lines in a circle. The problem was published by Van Ceulen in a pamphlet entitled *Solutie ende werckinghe op twee geometrische vraghen* (1584),[82] which was in turn intended as an answer to two questions posed by Willem Goudaen, a Haarlem arithmetic teacher, who had offered a reward to anyone who could solve the queries. Independently of each other, Nicolaas Petri and Ludolf van Ceulen succeeded. But Goudaen refused to pay the reward, citing all kinds of phoney excuses. The mathematicians took revenge by having their solutions published by Cornelis Claesz. It is not clear whether Coignet ever received a response from Galileo.

---

[80]Drake (1978a, p. 13).

[81]*Bella intestina miserabilis nostae inferioris Germaniae adeo bonarum artium studie extinguerunt, ita quod vix apud nos aliquem invenies, qui his artibus et studiis favere videatur.*

[82]van Ceulen (1584), f6*r*–6*v*.

Chapter 6
Wine Gauging

6.1 Antwerp and the Wine Trade

In the port of Antwerp – as in any other port – the most commonly used container type for transporting goods was the barrel. The history of the barrel goes back to the first century B.C.: it was a Celtic invention that was soon adopted by the Romans. The barrel had advantages over other types of containers. It was lighter than an amphora of the same content and easier to stack for transportation than leather sacks. Moreover, the wood of the staves gave the liquid contained in the barrel a specific taste that was deemed pleasurable to the palate. The wine trade in Europe was given a decisive impetus by the early Church Fathers, who insisted on the use of wine during the Eucharist.

Excises, i.e. indirect taxes on various commodities, were a very important source of income to the cities that levied them. In some towns, excise duty accounted for up to 80 or even 90% of the public authorities' revenues.[1] Excises on wine, which were often the first to be levied, seem to have been introduced in the thirteenth century in most European cities.[2] In England, the royal penny was introduced before 1256.[3] It was levied on each barrel of wine imported into England. Excises in general were commonly used to fund religious institutions. In Ghent, Bruges and Damme, for example, at least part of the proceeds of the wine excise went to the Hospitallers of St John.[4]

---

[1]Sosson (1990, p. 200); Bruges: 1332–1333 89.3%; 1350–1351 81%; 1391–1392 92.3%; Damme average 73%; Louvain fourteenth-century average 90%.

[2]E.g. Cologne in 1206 (Pauly (1994, p. 13)), Lübeck between 1220 and 1226 (Bornhöft 1985, p. 29), Liège in 1231 (Chaineux 1981, p. 112), Mechlin in 1263 (Sosson 1990, p. 200), Damme in 1265 (Meskens et al. 1999, pp. 53–54; van Cauwenberghe 1986, p. 122), Brussels in 1295 (Lloyd 1982, p. 86).

[3]Lloyd (1982, p. 86.)

[4]Duvosquel (2009), Meskens et al. (1999) and Meskens (2005b).

A. Meskens, *Practical Mathematics in a Commercial Metropolis: Mathematical Life in Late 16th Century Antwerp*, Archimedes 31, DOI 10.1007/978-94-007-5721-9_6,
© Springer Science+Business Media Dordrecht 2013

Although there were certainly vineyards in the Low Countries, the climate – even during the warm Middle Ages – was not really suited for the cultivation of grapes. Local production was quantitatively and qualitatively inferior to that of wines from traditional vinicultural regions. Abbeys and monasteries would therefore acquire vineyards in the Rhine valley and France.[5]

It was not only wine and beer that was transported in barrels, but also honey, oil and fish. Being able to determine the volume contained, without pouring it out, was important to both vendor and purchaser. Indeed, when it came to excise duties, it was equally important to city authorities. Hence, by the thirteenth century, they had appointed wine gaugers to measure the contents of barrels. So it is not so strange after all that manuals were written and – at least in the sixteenth century – published on this topic. Very often the writers of such manuals were arithmetic teachers, some of whom also acted as wine gaugers. Again, as with the arithmetic books, we find many examples of gauging manuals in Italy, Austria, Germany and the Low Countries, and far fewer in England and France.[6]

In the sixteenth century, Antwerp served as a hub for the wine trade: Rhine and French wines were distributed across the Flemish and Dutch provinces,[7] while Mediterranean wines were shipped on to the Baltic states.[8] In the design of the canals in the new town, a waterway with quays was envisaged where French and Spanish ships could unload their barrels of oil, syrup or wine, and where vessels from the staple at Dordrecht could unload their cargo of Rhine wines.[9] By the mid-sixteenth century, the local market shares of the different wines were 50% Rhine, 25% French and 25% other.[10]

The beer market, on the other hand, was dominated by local[11] and Brabant brews. English, German and Baltic beers accounted for only 5–10% of the market.[12]

During the fourteenth century, an excise duty on wine and beer was introduced alongside excises on commodities such as butter, cheese, herring and salt, as well as on the milling of grain. The wine excise was farmed out with annually determined

---

[5]See Chaineux (1981). The Church of St Martin in Liège, for example, owned vineyards in Worms and Bonn.

[6]See Bockstaele (1970), Folkerts (1974, 2008), Meskens (1994a), Meskens et al. (1999) and Simi and Toti-Rigatelli (1993). The large number of known German wine gauging manuals is due to the intensive and incessant research by Menso Folkerts. Some treatises on wine gauging are not easily identified, especially if they are part of, for example, a geometry treatise. For instance, a French treatise on the gauging rod appears in Chuquet's *La Géometrie* (1484) (Chuquet and H. l'Huillier (ed.) 1979, pp. 421–430).

[7]Craeybeckx (1958, p. 12).

[8]Craeybeckx (1958, p. 98).

[9]Soly (1977, p. 207).

[10]Craeybeckx (1958, p. 12).

[11]Prior to that, local beer production had accounted for about 55% of the market. This proportion grew in the second half of the century, owing to, among other things, Gilbert van Schoonbeke's entrepreneurship in this sector (Soly 1968).

[12]Craeybeckx (1958, p. 98), Renard (1921) and Soly (1968, pp. 347 and 1188).

rent that was payable per month or even weekly.[13] The importance that local authorities attached to wine excises is quite apparent from the fact that no barrel unmarked by the tax collector could be sold in the city.[14]

The city council imposed a tax not only on the transportation of wine barrels, but also on the sale of alcohol in the city. To enforce this rule, controls were required, which were carried out by two offices, one for beer excise, the other for wine excise. Since the beer excise office was concerned primarily with controlling the city brewers, not with measuring the content of barrels,[15] the activities of the wine excise office are more enlightening about the mathematical practices of the day.

6.2 Collecting the Wine Excise

Antwerp's wine excise office generally employed two wine gaugers, who would be appointed for life. By comparison, Paris, the only city in Western Europe that was larger than Antwerp, had had six wine gaugers and six apprentices on its payroll since 1486.[16] Antwerp's statutory team of two gaugers would occasionally be extended to three, as in 1551, following the escalation of a conflict between the excise officers and the gaugers over the accuracy of measurements. Wine merchants complained about the situation to the mayor and asked him to appoint an additional wine gauger.[17] They even proposed a candidate: Peter van Aelst, a gauger from Nuremberg, whom they claimed to be very competent. After a contest with the two official gaugers, Van Aelst was appointment as the city's third gauger.

Wine gaugers were accompanied during the execution of their tasks by a so-called *vuytschryver*, who noted down the measurements on behalf of the excise officer (Figs. 6.1 and 6.2).[18] After measuring the barrel, the wine gauger would brand his mark onto the barrel so that, if his calculation was subsequently found to be incorrect, he could be held to account. Of course, even with two officials dealing with merchants, bribery was conceivable. Heavy fines were therefore imposed on anyone caught taking bribes. A wine gauger, for example, could expect to be fined at least 25 gl. or even face dismissal.

In addition to measuring full casks, wine gaugers also measured any unsold quantities of beer and wine at inns, in order to fairly calculate taxes due on consumption.[19] Wines in transit were taxed differently, while wines that had turned sour were exempt from excise. Moreover, wine for consumption at inns was taxed at

---

[13]SAA, T6, Accysboeck, f iij *r-v.*

[14]SAA, Pk917, p. 170.

[15]Masure (1986, p. 91).

[16]Portet (1993–1994, p. 469).

[17]SAA Pk1409, f8*r-v*.

[18]SAA Pk1409, f2*r*.

[19]SAA T521, Wijnaccijns.

Fig. 6.1 Wine gauger at work. The man at the front of the picture is constructing a quadratic gauging rod (Mennher 1565, MPM A 3589)

a higher rate than wine intended for private consumption.[20] The wine gaugers were required to visit the city's inns for measurements once per quarter. This meant they could be quite occupied. In 1576, the delivery of an instrument to Plantin by Michiel

[20]Masure (1986, p. 77).

Fig. 6.2 Frontispiece (detail) of Raets and Coignet (1580a) (EHC G16053)

Coignet was delayed because the latter was too busy inspecting inns.[21] Obviously, the fact that the gaugers had to ascertain the level of consumption at inns meant they had to possess the mathematical skills to measure the contents of a partially filled barrel.

It is not known precisely how much gaugers were paid. Payment methods may well have differed over the years. A *contrerolle* (control book) from 1593 indicates that they received four stivers a barrel,[22] with the total amount divided equally between the two. Some of the account books of the City mention the annual amounts paid to the gaugers. As one might expect, these amounts varied quite considerably over the years, with payments as low as 21 gl. in 1585 and as high as 148 gl. in 1579.[23] Evidently wine gauging did not provide a steady source of income.

In 1590, in an appeal to the mayor, Michiel Coignet, who at that time was the city's senior gauger and keeper of the control book, claimed that his earnings from gauging were so small and paltry that it was impossible for him to continue to serve on the same basis.[24] He repeated his grievance in 1593 and 1595 before

[21]Letter from C. Plantin to Benito Arias Montanus. See Rooses and Denucé (1968, V, pp. 106–108), letter no. 692.

[22]SAA Pk1409, f43*r*.

[23]Meskens (1994b, p. 145).

[24]SAA Pk1409, f40*r*.

eventually resigning from his position as wine gauger in 1596.[25] In April 1593, Coignet compared his earnings of 180 gl. with those of the *vuytschryver*, who received 200 gl., and the bookkeeper at the excise office, who earned 400–500 gl. On this basis, he requested that the city pay him 400 gl. a year, but the authorities merely granted him a one-off supplementary payment of 100 gl.[26]

Doubts over the competence of the city's gaugers would persist over the years. In 1591, when the deputy sheriff called into question Coignet's abilities, the latter immediately responded. He suggested that the content of the disputed barrels be ascertained by means of the so-called water test.[27] Only then would it be clear who was right: "the gauger in gauging or the deputy sheriff in his rudeness".[28]

In his letter of resignation of 1596, Michiel Coignet suggested the reinstitution of the pre-1585 practice of appointing a gauger only after a contest.[29] Among the reasons cited is the claim that some gaugers "put the volume of a barrel more than an *aam* higher than the real volume", to which Coignet added that "they were no masters at their profession".[30] His assertion that some gaugers were incompetent was shared by Maarten vanden Dycke: "some gaugers do not understand the fundamentals of arithmetic and do not know how to make an accurate multiplication, especially when large numbers are involved".[31]

In addition to the practical test, Coignet also proposed that prospective gaugers should take an examination in the type of measures used in the cities and countries with which Antwerp traded: the Rhineland, Majorca, Seville, the Barbary Coast, and the Dutch staple towns of Dordrecht and Middelburg. If they were unfamiliar with the measures used in the towns of origin, they should be able to accurately measure the volume of barrels using the Antwerp measure. Unfortunately Coignet did not specify which practical test he felt the gaugers ought to take, but he did add that he felt such examinations were necessary because merchants had lost faith in the gaugers' ability. This mistrust manifested itself in fact that, as soon as a merchant had purchased a barrel, he would call in a second gauger to check whether the certified volume was indeed the true volume.[32]

Rather remarkably, when their income was jeopardized by the possibility of the appointment of a third gauger, Peter de Cock and Gabriel van Bemel rejected

---

[25] SAA Pk672, f47*r-v*, Pk674, f101*r-v*, Pk676, f76*r-v*.

[26] SAA Pk672, f47*r-v*.

[27] I.e. measuring the capacity of a barrel by filling it with water using a measuring glass, "as is the custom in this city".

[28] SAA Pk669, f161*v*.

[29] The wine gauging file of the Antwerp City Archives contains manuscripts describing two instances where a new wine gauger was appointed after a competition. One of these contests was held in 1567; the other took place in an unspecified year, but most probably between 1564 and 1576. SAA Pk1409, f13–14 and f35*r*. For a detailed analysis, see Meskens (1994a).

[30] SAA Pk1409, f49*r*.

[31] vanden Dijcke (1600, p. 19).

[32] SAA Pk1409, f40*r*. It seems Coignet is referring here to gaugers who measured barrels on behalf of merchants when no official control was necessary.

Hier van volcht een Tafel.

1.	1	000
	2	414
	3	732
2.	4	000
	5	236
	6	449
	7	645
	8	828
3.	9	000
	10	162
	11	316
	12	464
	13	605
	14	741
	15	873
4.	16	000
	17	123
	18	242
	19	359
	20	472
	21	582
	22	690
	33	796
	24	900
5.	25	000
	26	99
	27	196
	28	291
	29	385
	30	477
	31	567
	32	657
	33	744
	34	831
	35	916
6.	36	000
	37	82
	38	164
	39	244
	40	324
	41	403
	42	480
	43	558
	44	634
	45	709
	46	783
	47	856
	48	928
7.	49	000
	50	71
	51	141
	52	211
	53	280
	54	348
	55	415
	56	482
	57	549
	58	616
	59	681
	60	746
	61	810
	62	874
	63	937
8.	64	000
	65	62
	66	124
	67	185
	68	246
	69	306

1	1	—	27	5	196
2	1	414	28	5	291
3	1	732	29	5	385
4	2	—	30	5	477
5	2	236	31	5	567
6	2	449	32	5	657
7	2	645	33	5	744
8	2	828	34	5	831
9	3	—	35	5	916
10	3	162	36	6	—
11	3	316	37	6	82
12	3	464	38	6	164
13	3	605	39	6	245
14	3	741	40	6	324
15	3	873	41	6	403
16	4	—	42	6	480
17	4	123	43	6	557
18	4	242	44	6	633
19	4	359	45	6	708
20	4	472	46	6	782
21	4	582	47	6	855
22	4	690	48	6	928
23	4	796	49	7	—
24	4	899	50	7	71
25	5	—	51	7	141
26	5	99	52	7	211

Fig. 6.3 Tables with square roots from Raets and Coignet (1580a) (under, EHC G16053) and Mennher (1565) (above, MPM A 3589). Notice, in both cases, the absence of 0 as a place holder, e.g. the square roots of 26, 37, 50. This indicates that the figures are seen as the nominator of a fraction with denominator 1,000. On the other hand, the printer has introduced a right margin, thus always placing the thousands in the third column

Coignet's arguments. So despite De Cock being a former apprentice to Coignet and Van Bemel having received his endorsement, neither were willing to have their competence tested in the manner Coignet had proposed.[33] In the end, a third gauger was appointed.[34]

[33]In the case of Gabriel van Bemel, there may have been some nepotism involved. He was married to Isabelle Gramaye (see p. 42), the daughter of Cornelis I Gramaye, who held a lease on the wine excise. Meskens (1993a, pp. 27–28).

[34]SAA Pk1409, f53*r*-*v*, Pk697, f129*v*, Meskens (1994a).

Fig. 6.4 Frontispiece (detail) of vanden Dijcke (1600) (MPM A 3290). Although this book deals with wine gauging, its author Martin vanden Dycke, a schoolmaster and notary, was never employed as an official wine gauger

6.3 Explaining the Measurement Method

The practice of measuring barrel contents is explained in various arithmetic books, even though, with the exception of Michiel Coignet, no arithmetic teacher in Antwerp seems also to have been an official wine gauger. In some cases, as in the books by Valentin Mennher (Fig. 6.1), the approach is explained as an example of applied arithmetic. In other books, such as those by Willem Raets (1580 – Fig. 6.2) and Martin vanden Dycke (Fig. 6.4), it constitutes the subject matter of a separate treatise.

Wine gaugers generally assumed that a barrel was cylindrical. If the diameter d and the length l of a cylinder are known, then the volume is given by

$$V = \frac{\pi}{4} d^2 l$$

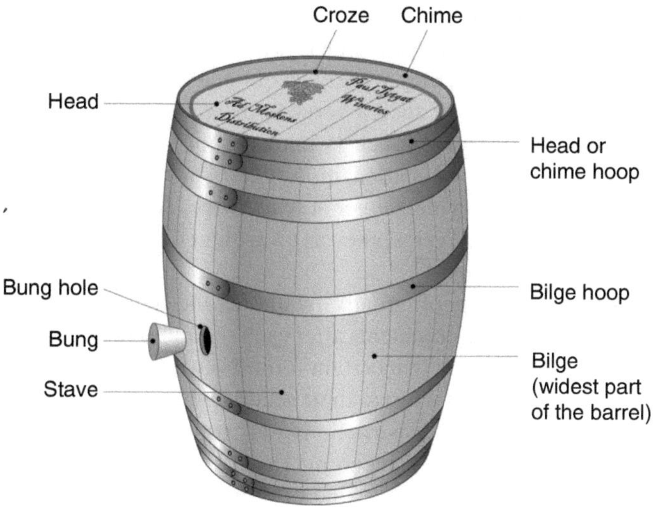

Fig. 6.5 © Paul Tytgat

If a cylinder of length l_0 and diameter d_o, and hence volume V_0 is taken as a standard, then the volume of any other cylinder can be expressed in terms of V_0. Taking $l = al_0$ and $d = bd_0$ then yields

$$V = ab^2 V_0$$

Since the latter formula does not involve π, the calculations are simplified. Although V_0 is arbitrary, the formula can be further simplified by assuming V_0 to be the unit volume used in the city.

It was further assumed that a cylinder of the same length and whose base area was a mean of the base areas at the heads and at the bulge of the barrel (i.e. at the widest diameter) had the same volume as the given barrel. Given the diameters at both heads and at the bulge, the mean base area is:

$$G = b^2 = \frac{1}{2}\left[\frac{1}{2}\left(d_{head1}^2 + d_{head2}^2\right) + d_{bulge}^2\right] \tag{1}$$

(For notations, see Fig. 6.5).

Setting $V_0 = 1$, the unit volume gives

$$V = aG$$

where $a = \dfrac{l}{l_0}$ as before.

To simplify the calculations, a quadratic scale was used. In the competitions organized by the city of Antwerp, the candidates were required to construct their gauge geometrically. In 1567, they were even obliged to make them on the spot.[35]

The construction and use of such a verge is the subject of Coignet's adaptation of Willem Raets's treatise. Not much is known about Willem Raets, except that he came from Maastricht and settled in Antwerp in the 1560s. He entered the competition to become city wine gauger at least twice. Although he was the author of an arithmetic manual, he never worked as an arithmetic teacher, but, together with his wife, was involved in trade.[36]

The presentation of the book greatly facilitates the reconstruction of the 1567 edition, as Coignet added his comments after each section. These notes include some points of criticism: Coignet felt that certain formulas were unacceptable, as they had been copied from other books without their accuracy having been checked.

The book begins with some constructions and propositions, such as the construction of perpendiculars and Thales' theorem. The construction of the quadratic rod is explained. First, one needs to draw a line of 4–5 in. This can be used to graduate the rod into a number of equal parts, called depth points. At the 1st, 2nd, 3rd, $\ldots$, nth division, the numbers 1, 4, 9, $\ldots$, n^2 are inscribed on the rod. For depth points that are not perfect squares, an iterative construction method is given. From Pythagoras' Theorem, we know that in a right-angled triangle with sides 1 and $\sqrt{n}$, the hypotenuse equals $\sqrt{n+1}$. In this fashion, all depth points can be constructed one after another (See Figs. 6.1, 6.2, 6.4).

The manual also gives a table showing the decimal parts above the previous depth point of a perfect square (see Fig. 6.3).[37]

While the unit for the depth was chosen arbitrarily, the graduation for length needed to be adapted if one wanted to express the volume in local units. To this end, one has to use a barrel with a known content V. Given that the equalized diameter is d_0^2, the number of divisions of the length l_0 of the barrel is given by $x = \frac{V}{d_0^2}$. Therefore, length l_0 has to be marked on the rod and divided into x equal parts. One such part is the unit length with which the length scale needs to be graduated.

Another method is given in a book by Ulrich Kerl [=Kern],[38] to which Coignet refers, adding that Kerl promises to describe an easy method, but actually explains it in a tentative manner. Coignet provides another way of graduating the rod, beginning

---

35 SAA Pk1409, f35.

36 SAA Pk640, f97*r-v*, 99*r*; Pk1409, f24–27 (1567), Cert 30, f251*v* (1569).

37 The table used to construct the rod gives (in modern terminology) the decimal parts above the previous perfect square depth point. The values of the square roots between 1 and 4 would be given as : 1 000, 2 414 3 732 4 000. This means that the second depth point should be $\frac{414}{1,000}$ units above the first.

38 See Leibowitz (1933, pp. 11ff. and 33ff.).

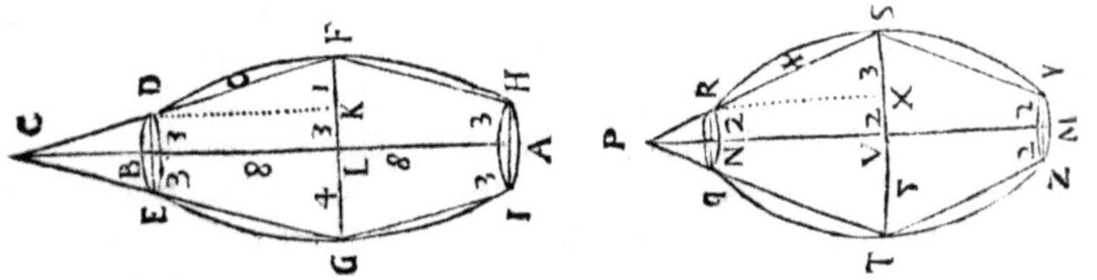

Fig. 6.6 Approximation by frusta of cones. Comparison between two cases in Mennher (1565) (MPM A 3589)

with its division into 24 equal parts. For a barrel with known volume V and length l, one finds the mediated diameter $d_0^2 = \frac{V}{l}$. The rod therefore has to have d_0^2 depth points (in his example $V = 1{,}464$ pints, $l = 30\frac{1}{2}$ and $d^2 = 48$). In most cases, d will not be an integer (here $d = \sqrt{48} = 6\frac{928}{1{,}000}$). Hence one can find the place of an integer diameter (e.g. $d_1 = 10$) using the rule of three. This will be the 100th depth point of a pint. Dividing into ten equal parts gives the 1st, 4th, ..., 81st depth point. Other points can be found in the same way as described above.

This approach was already criticized by Valentin Mennher in his 1565 book (fQ*q*v), where he points out that it only yields the correct result for similar barrels (Fig. 6.6). His argument runs as follows (see Fig. 6.6): consider two barrels, one with bulge 8 and equal heads 6 and another with bulge 10 and equal heads 4. The linear averaged diameter will in both cases be 7 $\left(= \frac{1}{2}\left(d_{bulge} + \frac{d_{head1}^2 + d_{head2}^2}{2}\right)\right)$. Thus if the two barrels have the same length, they will also hold the same volume, when the usual procedure, "in the way of the gaugers", is followed. Mennher argues that this is not the case, but incorrectly assumes that the frusta of cones formed by the bulge and the head in both cases hold the same volume. Mennher goes on to argue that the stave of the second barrel is more rounded than that of the first. Therefore it is longer and hence the second barrel holds a larger volume. The reasoning is somewhat flawed, since the volumes of the two frusta are not equal. Moreover, since the lengths of the lines along the surfaces of the two frusta are not the same, nothing definite can be said about the volumes of the respective shells.

In the sixteenth century, in order to avoid multiplication of length and depth, one would use a so-called change rod. The first known Dutch manual explaining the construction of a change rod was the book by Willem Raets edited by Michiel Coignet.[39] On either side of the change rod, one or more tables were engraved in

[39]As the 1580 edition is an annotated reprint of the 1567 edition, it is likely that this 1567 book was actually the first Dutch manual in which the change rod was described. Although we know the book was published, there is no known extant copy. The 1580 edition was reprinted in 1597.

such a way that the volume could be read from it upon taking the measurements (Fig. 6.7). First, the gauger has to calculate the equalized diameter. He measures the diameters of both bottoms and calculates their arithmetic mean in order to find the averaged bottom. He then measures the depth at the bulge and calculates the arithmetic mean of this depth and the averaged bottom. If the latter mean is an integer value, he can turn his gauge towards the correct *wissel* ('change table', hence change rod). With this *wissel*, the length is measured and the volume can immediately be read.

There was no consensus on which method is best applied in calculating the volume of a partially filled barrel. In fact, some manuals did not even mention a formula for dealing with this problem. For "in measuring the *wannicheyt* (litt. emptiness) lies the secret of the Gauge, yet nowhere do we find the right material",[40] i.e. the gauger's competence is most apparent when dealing with partially filled barrels, yet one cannot find a good description of this practice.

In the manual by Raets, we find a formula that, in anachronistic notation, looks as follows:

$$G = \frac{1}{2}\left[\frac{1}{2}\left(\frac{d_{bulge}^2 + d_{depth}^2}{2} + \frac{d_{head1}^2 + d_{head2}^2}{2}\right) + d_{depth}^2\right]$$

$$V = aGV_0$$

where "depth" denotes the depth of the liquid in the horizontal barrel at the bunghole, V_0 the volume of the gauging standard, and a the length relative to the gauging standard of length.

Coignet, the editor of Raets's book, found this formula unsatisfactory and asserted that authors of wine gauging books "all write the same, one copying the other without checking; for the test shows this formula to be false".[41] Calculations have indeed shown that the relative error generated when using this formula is quite large.[42] Coignet noted that certain Italian tables gave the volume of a barrel relative to the depth of the liquid. They were, however, based on differently shaped barrels from the ones used in the Low Countries and so were of no use. He further claimed that, in 1573, he had found "the General Rule, using the Art of Geometry"[43] which solved the problem of determining the volume of liquid in partially filled barrels. Unfortunately he did not give his rule, "because it takes more instruction than can

---

[40]Raets and Coignet (1580a), f Q ij *r*.

[41]Raets and Coignet (1580a), fQ*v*–Qij*r*.

[42]Meskens (1994a, pp. 145–146).

[43]Raets and Coignet (1580a), fQij*v*.

be given here". We may surmise that the additional instruction required involved an early version of the sector or, more likely, his pantometric rule (see Chap. 7 Sect. 7.2).

The book does not mention the cubic gauging rod, a kind of rod that seems to have been in use primarily in Austria.[44] The use of this rod is based on the property that for two similar bodies B_1 and B_2, which have known volumes V_1 and V_2 and diameters d_1 and d_2 resp., the ratio of the volumes is equal to the ratio of the cubes of the diameters: $\frac{V_1}{V_2} = \frac{d_1^3}{d_2^3}$.

This means that the verge has to be graduated similarly to the quadratic rod, with the numbers 1, 8, 27, …,n^3th indicated at the 1st, 2nd, …, nth point. Other points, in between the integer third roots, need to be constructed.

To make a measurement, the rod is inserted diagonally through the bung hole towards the lowest part of a bottom. With the correct graduation, the volume can then, be read immediately. None of the sixteenth-century Dutch manuals on gauging mention this method. As with the quadratic rod, the cubic rod can be constructed geometrically or algebraically. In the seventeenth century, tables of cubic roots were drawn up (e.g. by Sybrandt Cardinael[45]) for this very purpose.

In a manuscript,[46] Coignet deals with the construction of $\sqrt[3]{2}$, first referring to the story that the idol Apollo in Delos had asked to double the cubic altar (f3*r*). He asserts that no geometric solution to the problem (i.e. with ruler and compass) have yet been found, only mechanical ones (i.e. involving devices with moving parts).

The constructions which Coignet describes are all so-called *neusis* constructions. One of these is particularly interesting in the context at hand. Coignet's instrument (the principle of which he attributes to Plato) consists of a carpenter's square, *FGI,* along one of the legs of which, *GI*, a perpendicular transversary, *IK,* can slide (see Fig. 6.7). Two perpendicular lines, one horizontal, the other vertical, have intersection *O*. Choose a point *A* on the horizontal and use *OA* as a unit to graduate the lower vertical. Put the instrument on the drawing and let *GF* pass through *A*. Slide *IK* along *GI*, such that it passes through $D = V_2$ (or V_n) on the lower vertical, while turning the instrument until *G* is on the upper vertical and *I* on the left horizontal. The length of *OB* now equals $\sqrt[3]{2}$ (or $\sqrt[3]{n}$).

In a similar way, the other cubic roots can be constructed. "And this practice allows one to determine the cubic depth points on the rod, provided that one knows the first diameter *OA*" (f5*r*).

Coignet also describes his own invention to construct $\sqrt[3]{2}$, which is a variation on known methods (see Fig. 6.8).

In his manuals on the sector (see Sect. 7.2), Coignet on various occasions describes how one can find a barrel that has a similar shape and a volume equal

---

[44]Folkerts (1974, p. 31).

[45]Sitters (2008, p. 536 ff.).

[46]Coignet (1576–77)

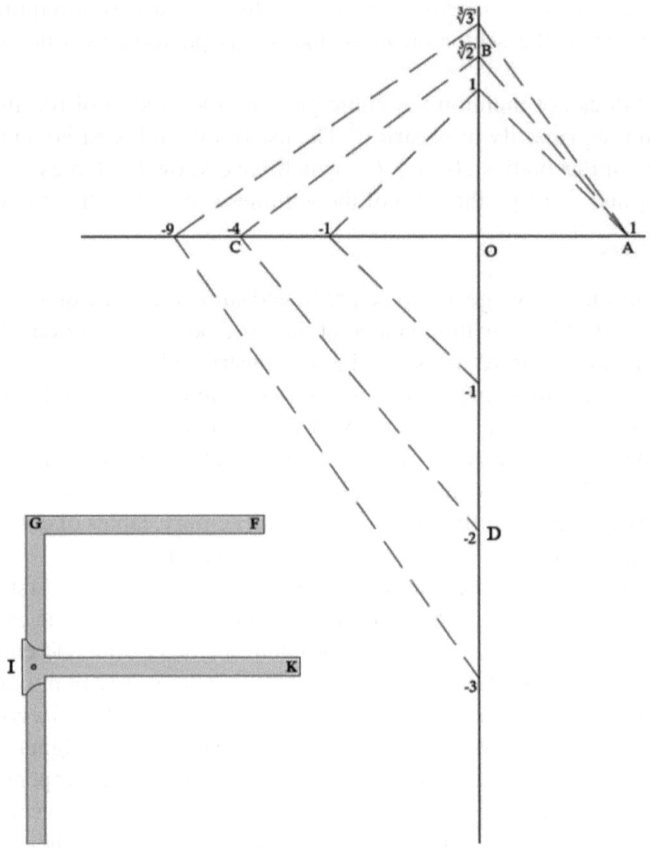

Fig. 6.7 Instrument for the construction of the third root (after a drawing in Coignet (1576–77)). This can easily be seen as $\Delta AOB \sim \Delta BOC \sim \Delta COD$
we have: $\frac{AO}{OB} = \frac{BO}{OC} = \frac{CO}{OD}$
now $OD = 2OA$
whence $\frac{AO}{OB} = \frac{BO}{OC} = \frac{CO}{2OA}$
from which, after some manipulations
$OB^3 = 2OA^3$ or $OB = \sqrt[3]{2}OA$,
since $OA = 1$ it follows that $OB = \sqrt[3]{2}$.
© Paul Tytgat

to the content of a partially filled barrel (Fig. 6.7). First one has to calculate the equalized diameter. With this diameter, the virtual cylinder with the same volume is constructed and the height h of this cylinder is determined.[47]

[47] Although he does not explain how to do this, the drawing suggests that the procedure is as follows: if d_m is the equalized diameter, d_b the diameter at the bulge, then $h_{\{cil\}} = h_{\{barrel\}} - \frac{d_b - d_m}{2}$.

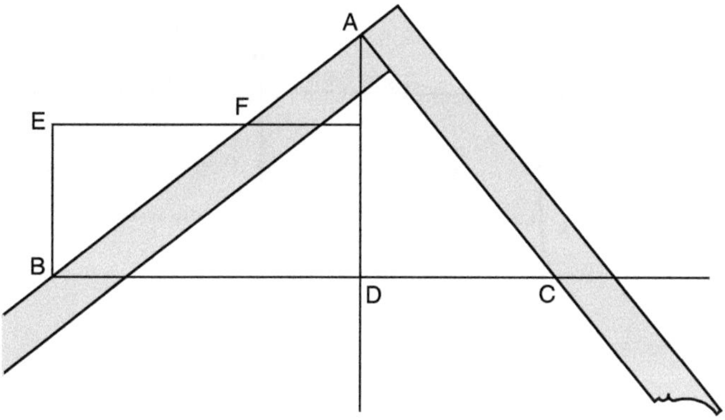

Fig. 6.8 Draw a rectangle with $2EB = BD$. With a carpenter's square (gnomon) the angle of which is on the vertical and going through A, such that $EF = DC$, then $EF = \sqrt[3]{2}EB$.
Indeed $\Delta EBF \sim \Delta DAB \sim \Delta DCA$.
Whence $2EB^2 = AD.EF$ and $2.EB.DC = AD^2$
$\Rightarrow 2EB^3 = EF^3$.
Coignet also describes the method in some of his manuscripts and in his book on the sector. Here he uses the sector folded out into perpendiculars instead of the carpenter's square.
© Paul Tytgat

Now draw a circle with diameter d_m, indicate the height h and draw a horizontal chord through this point. The sector allows equations of the type $\frac{f(x)}{f(y)} = \frac{a}{b}$ to be solved for any of the four parameters. Here the area of the part of the circle determined by the chord relative to the total area is calculated by using the function scale "D. cir." (circle division scale).[48] Suppose the area of the circle is S_C and the area of the empty part S_d, then $S = S_C - S_d$ is the surface area of the part filled with fluid. Using the line of solids (which is a cubic root scale), the dimensions of the new cask can be found. If l is the length of the first barrel, then putting this between the S_C-markings, the required length can be found between the S-markings.

Obviously in performing this procedure, Coignet also "solves" the problem of determining the content of a partially filled barrel.

In the second quarter of the seventeenth century, the sector, with a scale for "*wannicheyt*", would indeed be proposed as a gauger's tool by Adriaan Metius.[49] Coignet's sector procedure attracted the attention of, among others, Johannes

[48]In fact, one needs to determine $\frac{V_C}{V}$, but since $V_C = S_C l$ and $V = Sl$, this effectively amounts to calculating $\frac{S_C}{S}$.

[49]Metius (1623, 1626).

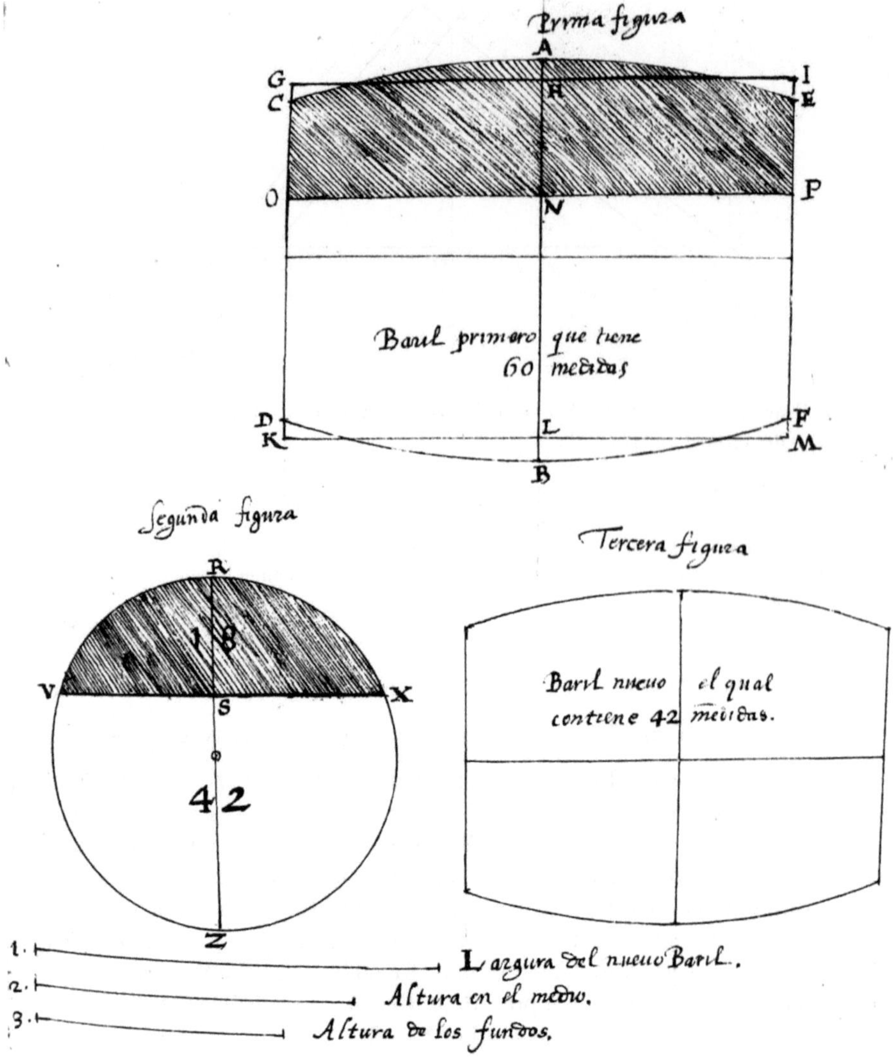

Fig. 6.9 Converting a partially filled barrel into a barrel that is completely filled (Coignet 1618b, EHC 520935)

Kepler, who refers to it in both *Stereometria* and *Messekunst*.[50] Kepler would elevate the theory of wine gauging to a higher level, by the incorporation of notions of stereometry and infinitesimal calculus.

[50]Kepler (1938–, 9, pp. 128, 237, 264–265).

Chapter 7
Instrument Makers

Antwerp's instrument-making industry, to the extent that there was any such thing, was dominated by the Coignets. By the same token, they were among the foremost instrument makers of the Seventeen Provinces.[1]

Whereas the gold and silversmiths, of whom the instrument makers were a subgroup, were far fewer than the schoolmasters, their respective trades seem to have been affected by similar trends: the religious troubles and economic pressures convinced a considerable number to leave the country. In a first emigration wave, many instrument makers emigrated to England, particularly London.[2] Before 1570, no fewer than 44 gold and silversmiths emigrated to London.[3] Then, particularly after 1575, Frankfurt gained in popularity as a residence in exile.

As previously mentioned, not all of these emigrating craftsmen were driven out by religious strife. Some, including instrument makers, were lured by the potential financial rewards. Cornelis Vinck, for example, had been active in Antwerp for some time prior to 1599, but then moved to Naples to further his career.[4]

There was, however, also a small reverse movement, as foreign craftsmen moved into the Southern Netherlands to benefit from the Archdukes' financial support to the Counter-Reformation. One such craftsman was Arnold-Florent van Langren.

[1]On instruments and instrument makers in the Low Countries, see Bennett (1996), Van Damme and van Cleempoel (eds.) (1997) and van Cleempoel (2002).

[2]The first known English mathematical instruments date from the middle of the sixteenth century, around the time when many instrument makers from Antwerp and Louvain resettled across the Channel. It is sometimes said that English instrument-making began with Thomas Gemini, a craftsman who moved from Louvain to London in 1544 (Brown 1979, p. 1; Bryden 1992, pp. 301–303). Gemini is often thought of as Flemish, but he actually hailed from the small Walloon village of Lixhe, in the then prince-bishopric of Liège. He may have worked at Frisius's or Mercator's workshop for a while before leaving for England (Ward 1981, p. 49).

[3]Briels (1971, pp. 89–90).

[4]Zinner (1956, p. 570).

A. Meskens, *Practical Mathematics in a Commercial Metropolis: Mathematical Life in Late 16th Century Antwerp*, Archimedes 31, DOI 10.1007/978-94-007-5721-9_7,

© Springer Science+Business Media Dordrecht 2013

Originally hailing from Gelderland, the Van Langrens[5] had moved to Amsterdam before settling in Antwerp at the turn of the seventeenth century.[6] A decade later, in 1610/1611, Arnold-Florent moved on to Brussels.

The work of Arnold-Florent and his father Jacob-Florent is so similar that it is hard to distinguish between them. Among other items, the Van Langrens produced a celestial globe based on Tycho's observations and with his permission.[7] Arnold-Florent is known to have supplied the King of Denmark with two small celestial globes.[8] In 1609, he produced a terrestrial globe for the City of Antwerp, for which he received 120 gl.[9]

Some instrument makers, like Ottavio Pisani, ended up in Antwerp rather by chance. Pisani was born in Naples, the son of Giovanni Antonio. He settled in Antwerp in the 1600s. Ottavio seems to have been a bit of an eccentric.[10] The only instrument he is known to have made was an armillary sphere.[11] Pisani also appears to have taken an interest in optical instruments, especially binoculars, on the subject of which he corresponded with Kepler and Galileo. In 1613, he published a remarkable book: *Astrologia seu motus et loca siderum* (1613). It is a lavishly illustrated title, with huge volvelles for determining the positions of the planets (see Fig. 10.11). By the time of its publication, however, it was already outdated.

7.1 The Coignet Family as Instrument Makers

Frisius's workshop had such a profound influence on the art of instrument-making in the Low Countries that we may well speak of the "School of Frisius". The instruments produced by his nephews Gualterus and Regnerus are particularly well-known. And his student Mercator, although better known as a cartographer, also ran his own workshop. Other instrument makers associated with this school are Adriaan Zeelst, Adrian Descrolières and, to a lesser extent, Michel Piquer.[12]

---

[5]See Keuning (1956) for a family biography.

[6]van de Vijver (1977, p. 80).

[7]MPM M385 is a manuscript manual on the use of the celestial sphere. It is attributed to Ottavio Pisani by Denucé (1927, p. 219). However, in the copy of Denucé (1927), used as an inventory by the Museum in the 1980s, O. Van de Vijver s.j. ascertains that it should be attributed to Arnold-Florent van Langren, which is indeed more likely. For the design of a sphere the author claims to have based himself on the observations of Tycho Brahe "des propres mains duquel i'ay receu l'obseruation des estoiles fixes" (f23*v*). Since both authors knew Tycho, it cannot be inferred from this remark who authored the manuscript.

[8]Zinner (1956, p. 424).

[9]Prims (1931b, p. 187). SAA Rekening 1609, f337*v*.

[10]Favaro (1895–1896).

[11]Rasquin (1996, p. 79).

[12]For a detailed history of Louvain instrument makers with an analysis of their creations see van Cleempoel (2002).

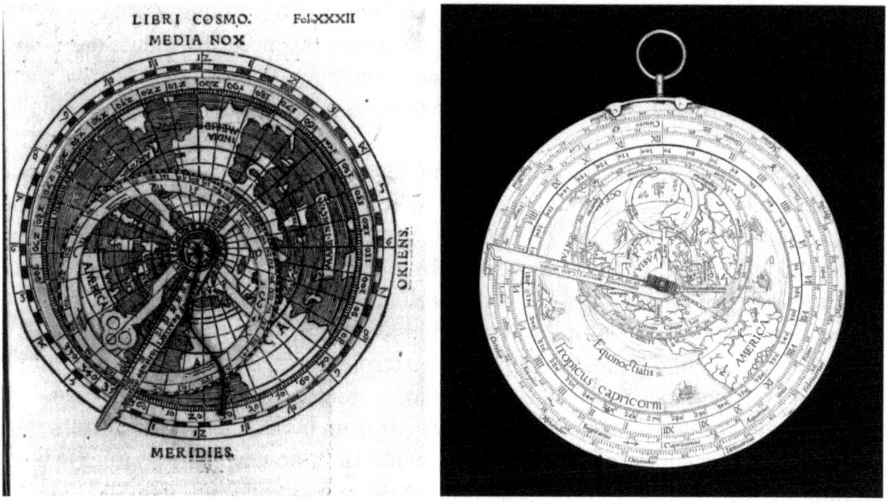

Fig. 7.1 A geographical astrolabe by Gillis Coignet (*right*, Museum of the History of Science (Oxford) 53211) and its description in Frisius and Apianus (1533) (*left*, MPM 4 186)

Despite a lack of material evidence, it is not unreasonable also to assume that the Coignets were associated with the school of Frisius. After all, Michiel Coignet employed Ferdinand Arsenius in his own workshop, and the latter is known to have cooperated with Gualterus Arsenius up until at least 1579.[13] His signature is first attested on an astrolabe dating from 1573. He seems to have moved to Antwerp after 1579, establishing a long standing collaboration with Michiel Coignet.

One of Gillis Coignet's astrolabes bears a striking resemblance to the astrolabe described by Gemma Frisius (see Fig. 7.1). A sundial by Gualterus Arsenius dating from 1558 would appear to be a mediocre copy of a similar instrument by Gillis.[14]

Only a few instruments manufactured in Gillis's workshop, all dating from the period 1557–1560, are known to us today.[15] In the 1540s, he provided Peter Frans, the city architect, with various instruments, including "an instrument of geometria" (valued at 12 gl.).[16] Possibly Frans required surveying instruments for improvement works on the city's fortifications.

---

[13]van Cleempoel (2002, p. 34).

[14]Michel (1947, p. 161). Not all instruments by Gillis were of a high quality. The astrolabe in the Museum of the History of Science in Oxford has been studied for its accuracy and compared with other astrolabes in the collection. Coignet's astrolabe has a rather large error, in the region of about 9,15′ per degree of arc. In fact, the error on Coignet's astrolabe is the largest of all astrolabes studied in Chapman (1983).

[15]J. Van Damme in Brokken (1994a, p. 158). The Gillis Coignet astrolabe of the British Museum is sometimes dated ca. 1580, but this is clearly incorrect. de la Solla Price (1955, p. 254). Misdating repeated in Zinner (1956) and Ward (1981). A sector signed Ægidius Coignet and dated 1561 was not made by Gillis, as will be demonstrated later (also V.A. Rasquin in van der Stock 1993, p. 299).

[16]SAA Pk712, f49*v*.

At the time of the death of his father, Michiel was too young to take over the instrument-making workshop, or indeed any other business his father may have run. By 1572, however, Coignet was already constructing instruments of his own, including an astrolabe, apparently based on designs by Gemma Frisius and examples by Gualterus Arsenius.[17]

This may indicate that his widowed mother had kept the workshop open since his father's death, which would have been in accordance with the regulations of the Guild of silver and goldsmiths [18]: a widow was entitled to keep shop and to employ personnel until the moment of succession by a son, as a Master. Becoming a master was however not an easy matter. It required an initial unpaid apprenticeship of 6 (subsequently 4) years, followed by a period as an unpaid workman in the shop of a Master. The candidate had to create a masterpiece, after which he was required to pay a considerable sum to the Guild and to treat the members to a banquet.[19] Michiel only became a Master in 1589.[20] Apart from his mathematical instruments, there are no other surviving artefacts by Michiel. He is however known to have been commissioned by the *kolveniersgilde* (one of the five armed militia guilds) to make golden coins, for which they paid him 5 gl.[21] He was also commissioned by the guild to manufacture a new chalice for their altar, but this work was ultimately executed by Niklaas Huybrechts the Elder.[22]

Printers seem to have maintained close relationships with instrument makers, and they were commonly involved in sales of instruments. It is not clear whether they purchased instruments to sell on to their customers, or whether they acted as agents on behalf of the makers. Also, Willem Sylvius, the only Antwerp printer able to compete with Plantin, is known to have owned an astrolabe.[23] Plantin, for his part, purchased two astrolabes from Regnerus Arsenius in 1569.[24] In 1575, Plantin wrote to Benito Arias Montanus (Fig. 3.5 left) that he had commissioned Michiel Coignet and Walter (= Gualterus) Arsenius to engrave certain plates.[25] Coignet, however,

---

[17]Zinner (1956, p. 281), also van der Stock (1993).

[18]Schlugleit (1969, p. 63).

[19]Schlugleit (1969, pp. 61–62).

[20]Schlugleit (1936, pp. 18–19); van Hemeldonck (1987, no.378); SAA GA4487, f254*v*, 255*r* en 258*v*. According to van Hemeldonck he was admitted to the guild as a Master's eldest son on 30 January 1590.

[21]SAA GA4664 1591, f14*v*.

[22]Prims (1938b, p. 306); SAA GA4664 1591, f14*v*. Niklaas Huybrechts served as Dean of the guild of silver and gold smiths in 1564, 1582 and 1590. He also served as Dean of *kolveniersgilde*. He died in 1605 and was buried in St Paul's Church.

[23]See Marnef (s.d.).

[24]Rooses and Denucé (1968, 5, no.683) dd. 17 December 1575, no.687 dd. 17/27 December 1575 and no.692 dd. 7 January 1576; MPM Arch47, f156*r*; also van Cleempoel (1997a,b).

[25]Rooses and Denucé (1968, 5, no.692) dd. 7 January 1576, suggest these plates are the central plates of an astrolabe. van Cleempoel (1997a, p. 78), on the other hand, argues that the reference is to copper engravings used in the printing of Arias Montanus's book. Nothing is known for certain about the relationship between Coignet and Arias Montanus. Still, Montanus must have known

was continuously engaged in wine gauging and hence was unable to deliver the plates in time. Both Arsenius and Coignet were paid 3 gl. for their work.[26] Coignet received further commissions through Plantin and his successors. In 1596, he was paid 100 gl. as a cash advance for brass instruments that were to be delivered to Jan Boyer, a bookseller in Medina del Campo.[27]

In 1581 he supplied the City with two sundials, worth 12 gl., intended for the spire of the Cathedral of Our Lady.[28] It is likely that these dials were part of a set. One vertical sundial can show the sun's shadow only if the sun's path through the sky is smaller than half a circle, which is not the case in Summer. Therefore at least two dials were needed. The next year, he supplied another stone dial worth 6 gl.[29]

In fact, he must have supplied sun dials to many customers, as attested by Adrianus Romanus s.j. (also known as Adriaan van Roomen) [30]:

> Not mentioning the many timepieces (horologia) he has installed in Antwerp walls and which he has described in a very valuable book.

It is unclear precisely which book Romanus is referring to, though Coignet did describe the construction of sundials in his manuscripts on the use of the *pantometer* (see p. 118 – Fig. 7.2). His student Saminiati moreover wrote a book in which he included the necessary tables for installing sundials.[31]

Coignet's workshop must have enjoyed a good reputation. In 1586, Jean Moflin, chaplain to Philip II, sent him an equinoctial sundial[32] to repair. He also requested instructions for its use, since it was intended as a present to the King (see Fig. 7.3 for an example of Coignet's fine craftmanship).

Upon his death, his tools and some remaining instruments were sold, including two sectors, an astrolabe (15 gl.), a ring dial, surveyor's sticks (2 gl. 8 st.) and a magnetic compass. Some, though it is not specified which, were sold to the Jesuits. A *trotera* (possibly this is a corruption of diopter or dioptra) and some books on timepieces were sold separately for the considerable sum of 88 gl. On the other hand, a sum of 18 gl. was due to Ferdinand Arsenius for the construction of pantometers (sectors).

Arsenius, because in 1570 he was recommended to the Duke of Alba for a lectureship at Louvain, together with Gemma Frisius (= Cornelius) and Mercator. There is also later evidence that the instruments of Frisius and Arsenius were used in the circle of friends of Montanus. Furthermore, Montanus is known to have possessed a collection of Antwerp engravings that interested Spanish artists (Rekers 1972, pp. 147 and 213).

[26]MPM Arch121, f106, cited in Denucé (1912–1913) 1, p. 240.

[27]MPM Arch21, f105v; Arch 126, f98 left; also Journal 1596, f35–36.

[28]SAA R20, f267*v*.

[29]SAA R23, f289*v*–290*r*.

[30]Van Roomen (Romanus) (1593), **j*v*.

[31]Saminiatus (1599).

[32]"ung horologe equinoctial et declinatoire avec aultre usaige". Rooses and Denucé (1968, 8, nr. 1170) dd. 4 November 1586.

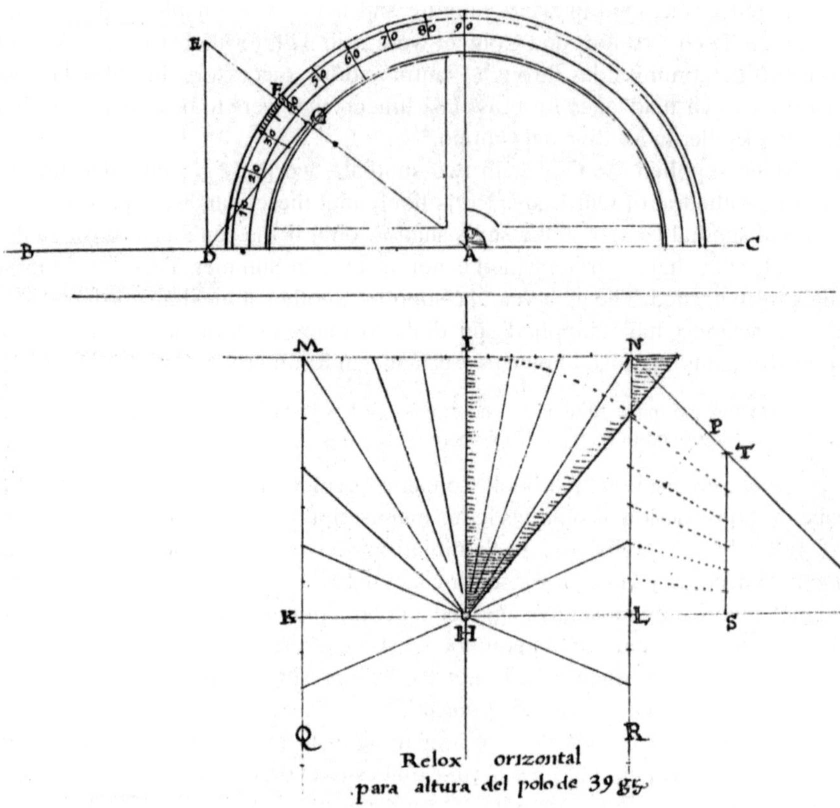

Fig. 7.2 Construction of a sun dial with the aid of a protractor in Coignet (1618b) (EHC 520935)

7.2 The Proportional Compass and the Sector

A *proportional compass* [33] consists of two slotted arms with sharp points at each end (see Fig. 7.4). In the slots of the arms is contained a cursor that serves as a pivot and can be held at the desired place by means of a screw. The pivot keeps the distances between the two points at each end in the same ratio, whatever their separation. It was a development from the X-shaped draughtsmen's instruments, which were fixed at ratios 2:1, 3:1, and so on for wider and narrower points. The

[33] As there is confusion about the proper terminology, we use that proposed by Stillman Drake in his book on Galileo's sector. What is referred to in English as a *sector* is known in French as a *compas de proportion*, in German as a *Proporzionalzirkel* and in Dutch as a *proportionaalpasser*. Confusingly, the English term *proportional compass* translates as *compas de réduction* in French, *Reduktionszirkel* in German and *reductiepasser* in Dutch.

On these instruments, see Schneider (1970), Drake (1978b), Meskens (1997a), Camerota (2000).

Fig. 7.3 Polyhedral sundial by Michiel Coignet (private collection).
Time can be told by the progress of a shadow during the day, which is measured by sundials. The style of a sundial is the time-telling edge of the gnomon (a thin rod or a sharp-edged element). The tip of its shadow traces out a conic section, which is the intersection of the cone traced out by the light rays and the (locally) horizontal surface of the Earth. This conic section changes with the seasons as a result of the Sun's declination. Sundials may be vertical, horizontal or reclining, the position of its style being determined by the latitude of its position. This polyhedral sundial combines all three types

proportional compass is said to have been designed prior to 1575 by Federico Commandino of Urbino at the behest of Bartolomeo Eustachio. In Coignet's 1618 manuscript, its invention is attributed to Michelangelo Buonarotti, the painter of the Sistine Chapel. The proportional compass usually has two sets of scales. One shows the number of times the distance between the narrow pair of points will fit between the wider pair. The other gives the number of sides of a regular polygon such that the wide pair gives the diameter of the circumscribed circle, while the narrow pair gives the length of the side of the polygon.

Another drawing instrument is known as a *reduction compass*. Invented by Fabrizio Mordente before 1567,[34] a reduction compass consists of two pivoting arms that lie parallel to a table supported by a point beneath the pivot and a pair of points at its extremities (see Fig. 7.5). A second pair of points runs on cursors

[34] Drake (1978b, p. 12) and Camerota (2000).

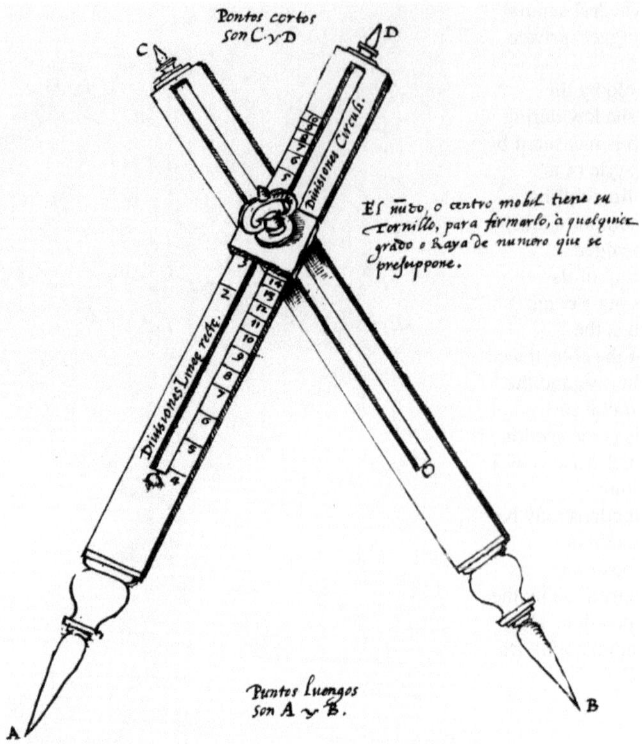

Fig. 7.4 Proportional compass (Coignet 1618b, f35*v*, *Prologo*, EHC 520935), Coignet attributes its invention to Michelangelo

and can be set to divide the arms in any desired proportion. A reduction compass generally carries no scales.

A *sector* consists of two pivoting arms with engraved scales, which can be used for computations. However, these computations are more easily performed than explained. The sector is based on the principle that the corresponding sides of similar triangles are proportional (in practice these amount to isosceles triangles). The scales on the arms are copies of a desired function f, e.g. the square roots, sines or tangents.

The principle of engraving these scales on the legs is the same as graduating the quadratic or cubic rod. If f represents the function square root, the numbers 1, 4, 9, …, n^2 are engraved at lengths 1, 2, 3, …, n from the pivot, as on a quadratic rod (see p. 106). Using the sector, the equation $\dfrac{f(x_1)}{f(x_2)} = \dfrac{a}{b}$ can be solved for any of the letters x_1, x_2, a, b provided the other three numbers are known (see Fig. 7.6).

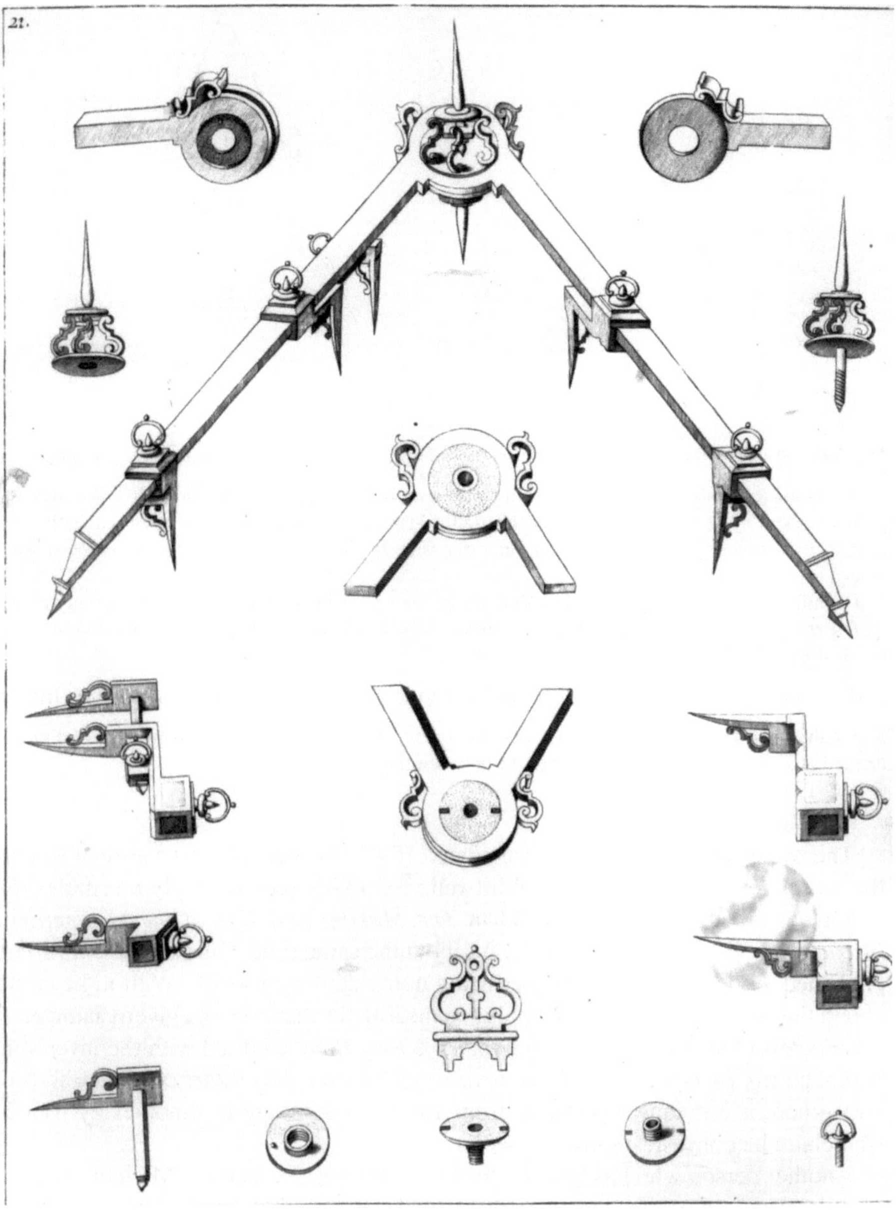

Fig. 7.5 Reduction compass with its parts (Mordente and Mordente (1591), KBR VI 9651 D)

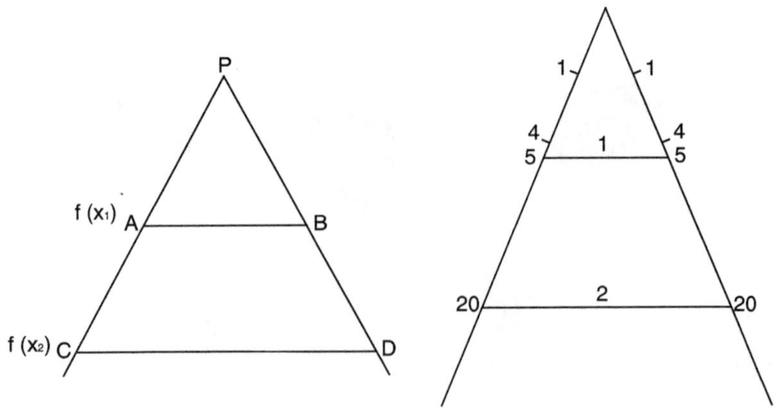

Fig. 7.6 To solve the equation $\frac{f(x_1)}{f(x_2)} = \frac{a}{b}$, one proceeds as follows. Suppose x_1 is unknown (solution analogously for x_2). Draw a line with length b. Open the sector in such a way that the points marked x_2 fit between the line with length b. Take a pair of dividers and open it at a distance a. Find two markings x_1 on the sector such that they fit the line with length a. x_1 is the desired value.
If a is unknown, draw a line with length b. Open the sector in such a way that the points marked x_2 fit between the line with length b. The length of the line between the points with markings x_1 is the desired value a.
In the second figure, an example is given of how to solve $\frac{\sqrt{5}}{\sqrt{x}} = \frac{1}{2}$ using this procedure. Draw a line with length 1, open the sector such that the line fits between markings 5. Using a set of dividers opened at 2, we find that the extremities fit between the markings 20.
© Paul Tytgat

The two earliest known sectors are dated 1597. One bears the signature of Robert Beckit, the other that of Charles Whitwell.[35] Beckit's sector closely resembles the sector described by Thomas Hood in *The Making and Use of the Geometrical Instrument, Called a Sector* (1598). At about the same time, Guidobaldo del Monte published a book in which he described a non-calculating sector. With it, he could obtain the length of the sides of a polygon inscribed in a circle of a given diameter. It was however Galileo Galilei who was, for a long time, credited with the invention. Although his *Le Operazione di compasso geometrico et militari* only appeared in 1606, there is a manuscript dating from 1599 and there are testimonies by friends who claim he constructed his first sector in 1598.

Another person who has been dubbed the father of the sector is Michiel Coignet. The descriptions in his manuscripts indeed suggest that he developed such an instrument. Coignet's manuscripts can be divided into three categories: the first are those in which he describes the *reigle platte*, the second concern his adaptations of Mordente's sector, and the third describe a proper sector. On the basis of these writings, one can reconstruct the development of his sector.

[35]Turner (1983, p. 103) and Bulletin of the Scientific Instrument Society no. 38(1993), 28.

A course delivered to quartermaster Thomas Franquin[36] illustrates the *reigle platte*, while other parts of the manuscript deal with trigonometry and applied trigonometry, in particular in relation to fortifications. The section on the *reigle platte* (the plane ruler) dates from 1610. At that time, Coignet had already adapted Mordente's compass. With the *reigle platte*, all operations performed with a sector can be made on paper:

> Le requis est ainsij une ligne laquelle est aucunes fois, ou le diametre ou le costé dune figure plane, ou dung corps que on cherche. Operation ou Inquisition du requis se faict presq generalement en ceste manier ensijante. Tirez une ligne droicte aplaijsir, laquelle cij apres se nomera la ligne Fondamentale, a l'extremite de laquelle lon marquera vn point lequel s'apellera le point centrale, Dauantaige en la susditte ligne fondamentale, il faut notter doiz le point centrale. Vers lautre extremite deux longueurs pris hors des diuisions de la reigle pantometre (en suite la teneur de la proposition) desquelz le premier representera le donne, et interual de ces deux longueurs deux arcs de cercle au premier desquelz (qui signifie le donne) de marquera la longueur de la ligne donnee, et par lextremite de Iceluij arc tirez au point centrale une ligne droite, coupant le second arc au point requis et la ligne soubtendue de cest arc donnera le requis. (f40*v*)
>
> *Suppose we want to solve* $\frac{a}{x} = \frac{f(x_1)}{f(x_2)}$ *in which* x *is the unknown then (anachronistically in modern notation to avoid cumbersome digressions)*
> *Draw a ray k with initial point A*
> *Measure the length* d_1 *of* $[0, f(x_1)]$ *on the pantometer*
> *Construct* B_1 *on* k, *such that* $\|AB_1\| = d_1$
> *Repeat for the length* d_2 *of* $[0, f(x_2)]$ *and construct* B_2 *on* k, *such that* $\|AB_2\| = d_2$
> *With A as centre draw circles with radii* d_1 *and* d_2 *respectively, i.e. circles through* B_1 *and* B_2 *respectively.*
> *Find the point* C_1 *on the circle through* B_1 *such that the length of the chord* B_1C_1 *equals* a $(\|B_1C_1\| = a)$
> *Draw a ray* l *with initial point A and through* C_1.
> *Call the point of intersection of this ray* l *with the circle through* B_2, C_2.
> *The length of the chord* B_2C_2 *equals the unknown* x $(\|B_2C_2\| = x)$

The *reigle platte* carries 12 scales (see Figs. 7.8 and 7.9) lines of equal parts (I), chords (II), the side of an inscribed polygon (III), side of a polygon with area equal to another polygon (IV), sines (V), tangents (VI). Scale VII is a line of areas with which the length of the side of a polygon is given, the area of which is a multiple of a given polygon. Obviously this can also be used to calculate square roots. Scale VIII is a similar scale for solids, with which cube roots can be calculated. Scale IX is used to calculate the area of a segment of a circle. Scale X is the spatial equivalent, with which the volume of a segment of a sphere can be calculated. Scale XI is analogous to scale IV. Given a Platonic solid, this line allows one to calculate the length of the edges of one of the other Platonic solids, with a volume equal to that of the given Platonic solid.

The oldest known Coignet manuscript on Mordente's reduction compass dates from 1608 and is kept at the Biblioteca Estense in Modena.[37] Another, partly printed, manuscript, which is in the same library, has neither date nor printer's mark,

---

[36]The part on the calculation of sines was edited and published in Bosmans (1900).

[37]Coignet (1608), other manuscripts with comparable content are Coignet (1616, s.d.a).

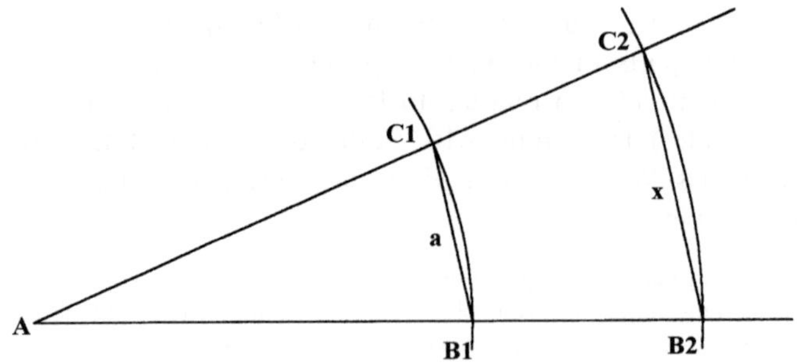

Fig. 7.7 After a figure in Coignet (1610–1612)

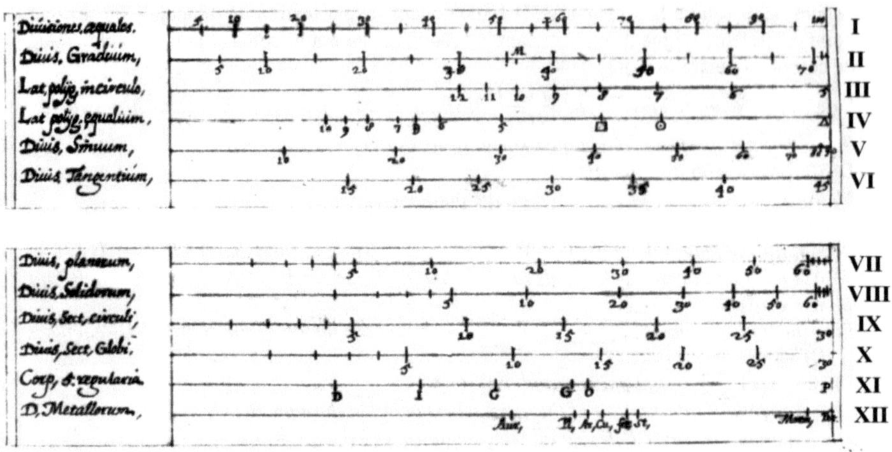

Fig. 7.8 The *reigle platte* (Coignet (1610–1612), KBR Hs II769)

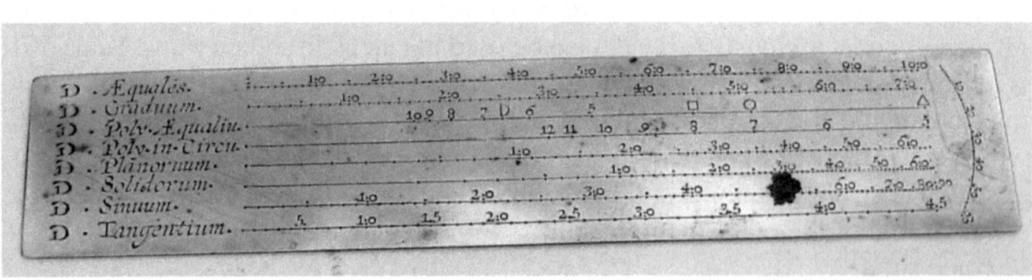

Fig. 7.9 A Coignet type *reigle platte* sold by Tesseract (Hastings-on-Hudson, NY) in 2008

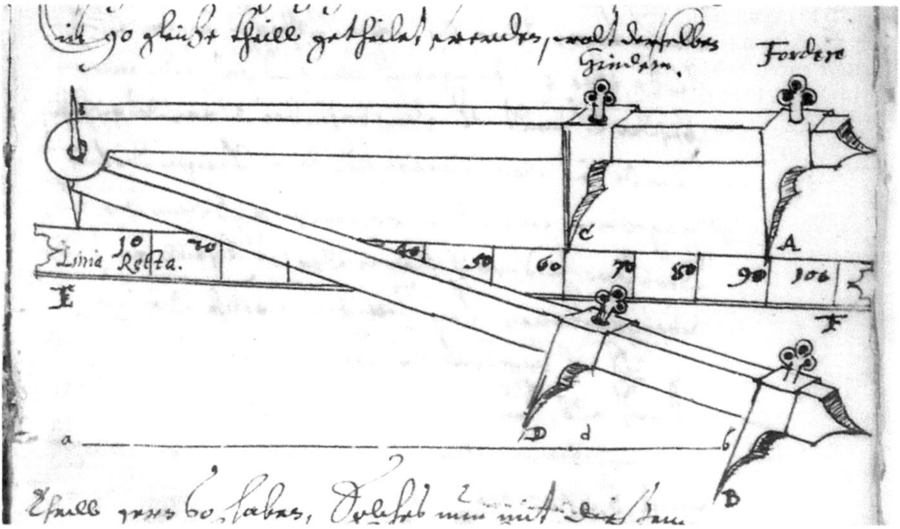

Fig. 7.10 Mordente's reduction compass in Coignet's version (Coignet (s.d.b), University Library Wrocław Ms R461)

but according to an annotation it was printed in Antwerp in 1608. The latter has printed text with hand-drawn figures.[38] Both are in Italian. These manuscripts are very important, because they show that Coignet had added a feature to the compass: a graduated rule. On Coignet's rule, we find a line of equal parts, a line of chords, and Guidobaldo's scale. On the line of equal parts, a scale was added with which it was possible to convert a circle into a square with the same area, and a sphere into a cube with the same volume. There was also a scale of inverse numbers, and the Modena manuscript[39] suggests that there were still other scales. At a certain point, three further rules were added, as the Wrocław manuscript attests (Fig. 7.10).[40] Each of these rules carried three scales, making this reduction compass equivalent to the sector. If any one rule was used in conjunction with the compass, the operations performed would have been the same as with a sector. In Coignet's posthumous book *La Géometrie* (1626),[41] it is asserted about the scales that

> [Coignet], tres-grand amy de Fabrice Mordente, autheur du Compas, auquel mesme por pus grande facilité il a adiousté deux regles que tu y vois
> *[], very good friend of Fabrizio Mordente, inventor of the compass, to which, for greater ease of use, he added the two rules you see here.*[42]

This may refer either to the pantometric rule (*reigle platte*) or to a sector.

[38]Coignet (1608).

[39]Coignet (1608).

[40]Coignet (1616).

[41]Coignet and P.G.S. (edit.) (1626), introduction.

[42]Coignet and P.G.S. (edit.) (1626, p. 3) of the Introduction.

Fig. 7.11 Euclid, using a Coignet-like compass to measure an angle (Marolois (1614–1615), EHC G 48942)

Coignet may have met Mordente on a number of occasions. The latter accompanied his patron, Allessandro Farnese, on his campaign in the Low Countries, so it is possible he came face to face with Coignet in the aftermath of the Sack of Antwerp in 1585.[43] Mordente reentered the service of the Duke in the 1590s, which again yields opportunities for meetings between the two men. Unfortunately, though, there is no evidence to substantiate this theory further. Coignet may in any case have become acquainted with Mordente's compass in a number of other, indirect, ways. First, in 1584, Christopher Plantin published a book by Gaspare Mordente, Fabrizio's brother, that dealt with the reduction compass.[44] Second, in 1591, the Antwerp printer Philip Galle published *La Quadratura del Cerchio*.[45] Most of this 22-page book deals with constructions with which a circle can be transformed into a rectangle of equal area. The final two pages contain wonderful drawings of the reduction compass.

[43]Rose (1968, p. 69).

[44]One copy is preserved at the Biblioteca Vittorio Emmanuale in Rome, while a sixteenth-century manuscript copy is kept at the Staatsbibliothek in Munich, and a third is known to have been sold at auction. It is not surprising that so few copies are known, since the book was published on only 18 copies on behalf of Monsr. de Cruyninghe, who paid the considerable sum of 82 gl. 10 st. MPM Arch20, p. 205.

[45]Mordente and Mordente (1591).

Coignet may also have read Mordente's Italian book *Modo di trouare con l'Astrolabio, ò Quadrante, ò altro instromento [...],*[46] in which the reduction compass was described and which was published in Venice by P. Forlani. Coignet is known to have had a large library containing a sizeable collection of foreign books. In fact, this library and Coignet's knowledge of the scientific literature were already acknowledged by his contemporaries.[47]

Two manuscripts in Latin[48] would appear to provide the oldest known descriptions of a type of sector that can also be used as an observation instrument (the "surveying sector" – see Figs. 7.12, 7.13 and Fig. 7.11 for an allegorical depiction). Apart from some minor details, these manuscripts are identical and they were presumably copied from the same source. In the manuscript, Coignet describes a sector with scales relating to trigonometry (equal parts (I), sines (V) and chords (II)). This sector still bears some resemblance to Mordente's compass, and indeed some of the problems relating to surveying are solved in the same way. In the first three books of the manuscripts, Coignet solves problems taken from practical trigonometry that relate to finding distances, altitudes, carrying out triangulations, and the construction of fortifications.

In these manuscripts, the sector is also used as an angle-measuring device, sometimes with the aid of a magnetic compass.[49]

In the preface to the Oxford manuscript, Coignet mentions that he crafted the first example of this kind of sector for Archduke Albert and the second for Hercule Blancho (Biancho?) of Milan. The binding of the Paris manuscript bears the coat of arms of Archduke Albert, suggesting that it was this copy that was given to the former. If this was indeed the case, then it was produced no earlier than 1596, when Albert became Regent of the Southern Netherlands. This was of course the same year as Coignet entered into the service of the Hapsburg Court in Brussels.[50] The astronomical examples in the manuscripts provide a further clue to the dating of the manuscripts. The positional data for two stars are given for the year 1604.[51] And a few pages later, he writes [52]:

> Anno 1603 die 24 Augusti *investigaui* Antuerpię ante meridiem solis altitudinem
> *In the morning of 24 August 1603* I observed *the solar altitude in Antwerp*

---

[46]Mordente (1567).

[47]For example, in 1611 D. Antonini wrote to Galileo that *Diagna Astronomica* was not available and that *even* Coignet could not be convinced that the book existed. Antonini then advised Galileo to wait for the *Frankfurter Buchmesse*. Antonini (1588–1616) was an Italian mathematician and soldier who had studied with Cataldi and Galileo. In 1611–1612, he served in the Army of Flanders (see Drake 1978a).

[48]Coignet (ca. 1604,a).

[49]The measured angle can either be transferred directly onto a drawing board, or can be read from a graduated rule. On this rule is engraved a table of chords, for a circle with given radius. This radius is indicated on both legs of the compass. By putting the line of chords between these points, one can read the angle subtended by the legs.

[50]SAA Pk676, ff76*r-v*.

[51]Coignet (ca. 1604)f31*v*, Coignet (after 1604).

[52]Coignet (ca. 1604)f35*v*, Coignet (after 1604).

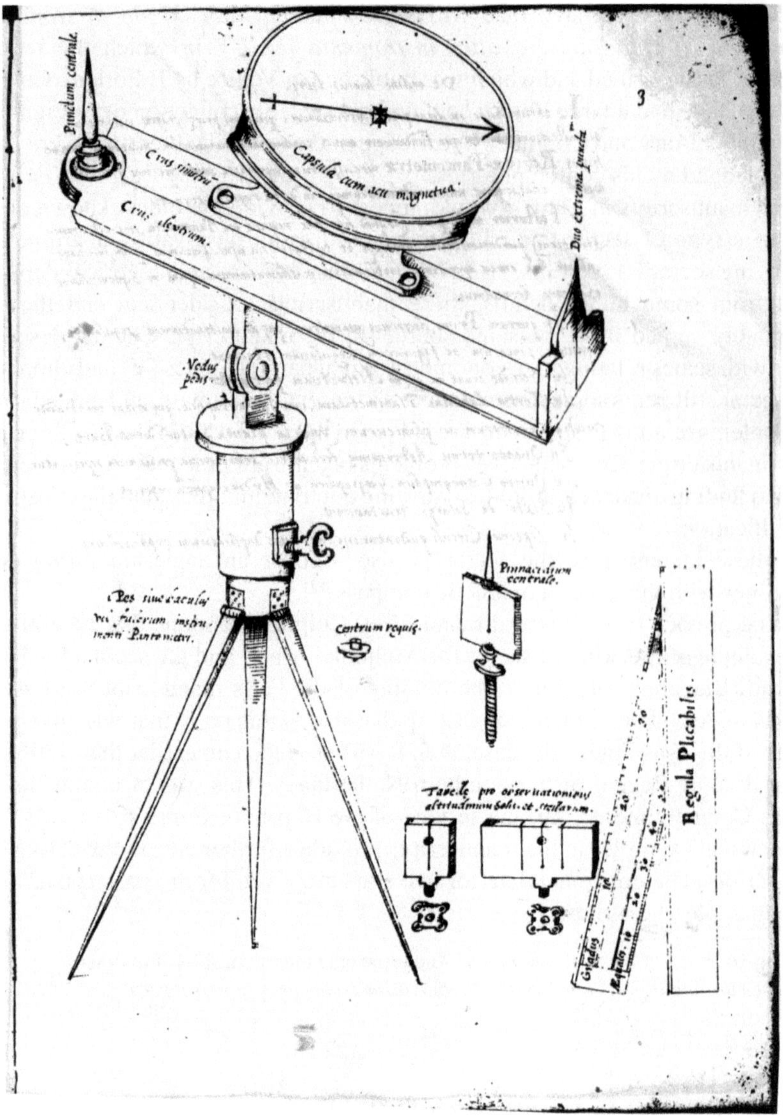

Fig. 7.12 Coignet's adaptation of Mordente's compass. It is fitted with a magnetic compass for surveying purposes (Coignet (ca. 1604), Bibliothèque nationale de France (Paris), J253)

These two references clearly indicate that the archetype manuscript was written around 1604.[53]

[53]In the 1610/1612 manuscript (Coignet (1610–1612)), Coignet gives 1613 values as an example. If he actually observed the solar altitude, this provides a date "post quem", and hence the 1604 example puts the earliest date in the Autumn of 1603.

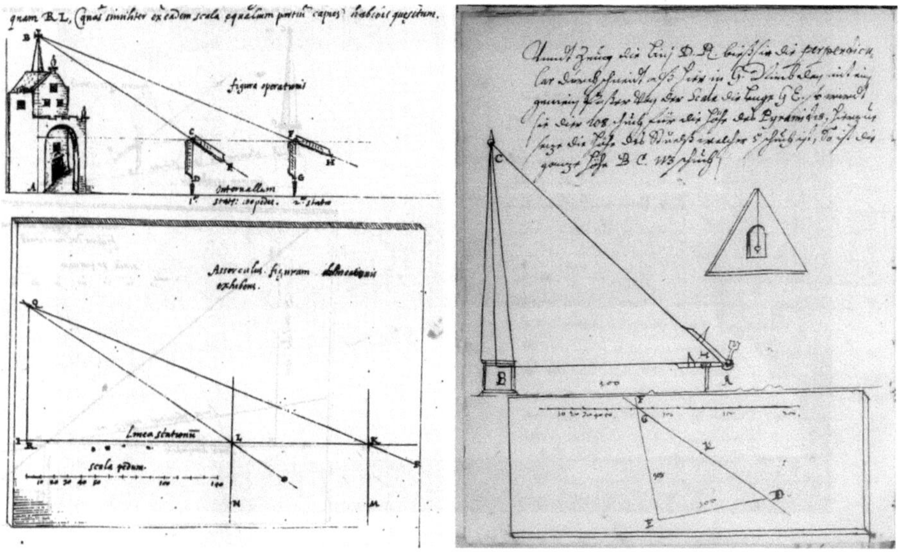

Fig. 7.13 The use of the Mordente-Coignet compass as a surveying instrument in Coignet (ca. 1604) (*left*, Bibliothèque nationale de France (Paris), J253) and Coignet (1616) (*right*, University Library Wrocław Ms R461)

The sector that is described in these manuscripts seems to have been developed from Mordente's reduction compass. It no longer carries moveable cursors, but has sighting vanes at the ends. It carries scales on its legs: a scale of equal parts, a scale of chords and a scale of sines. In this manuscript, the focus is on applications of geometry.

There is however an entirely different alternative explanation. Recently a sector has come to light carrying a scale of inverses on one side and a scale of the sides of an inscribed polygon on the other (see Fig. 7.14). Almost nothing is known about the instrument, making it very difficult to identify either date or origin. Some stylistic elements do point towards the first quarter of the seventeenth century and Flanders though (e.g. the circles around the pivoting point).[54]

If Coignet was indeed familiar with Gallucci's book *Della fabrica et uso di diversi stromenti*...,[55] which contains a description of this kind of sector, he would

[54]The instrument features in the painting *Allegory of Sight (Venus and Cupid in a Picture Gallery)* by Jan Brueghel the Younger (ca. 1660) in the Johnson Collection of the Philadelphia Museum of Art, and in the painting *Allegory of Sight (Venus and Cupid in a Picture Gallery with a view of Antwerp)* (ca. 1650–1670) in a private collection (Fig. 7.15), but the dates of these pictures render them useless in determining when these instruments were introduced in the Low Countries.

[55]Gallucci (1597) and Schneider (1970, p. 29). Given Coignet's large library and his knowledge of contemporary literature (see pp. 21 and 127) this assumption is not unreasonable.

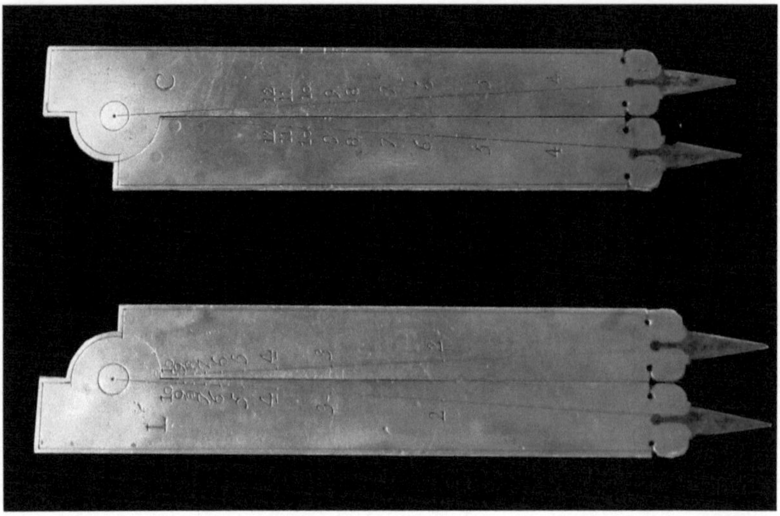

Fig. 7.14 A sector with a scale of inverses on one side and a scale of the sides of an inscribed polygon on the other (private collection)

most certainly have been able to make such an instrument by the turn of the century. Again, it would have been easy to add other scales.

Combining the ideas of Mordente, using the reduction compass as an observing tool, and the idea of the sector could then have yielded the new instrument that was the surveying sector. Unfortunately, as there is no written evidence to substantiate this theory, it remains entirely speculative.

This surveying sector was propagated as such by Samuel Marolois (ca. 1572–1627) in his book *La Géometrie* (1614).[56] Marolois was a Dutch mathematician who wrote on applications of geometry in designing fortifications, in architecture and in perspective. His books were reedited by Albert Girard (1595–1632). *La Géometrie* was published in French, Latin, Dutch and German. In this book, he describes the use of a Coignet-type sector in the surveying practice (see Fig. 7.16).

By the second decade of the seventeenth century, Coignet seems to have developed a pair of sectors that bear a superficial resemblance to Galileo's sector (see Figs. 7.17 and 7.18). Although the two share certain scales, others are quite different, suggesting that the tools may have been developed independently. The problems proposed in the manuscripts describing these sectors[57] are similar to those in the Oxford and Paris manuscripts.

[56]On Marolois and *La Géometrie* see van de Weyer (2011).

[57]Coignet (1618b, ca. 1610b).

Fig. 7.15 Jan Brueghel the Younger, *Allegory of Sight (Venus and Cupid in a Picture Gallery with a view of Antwerp)* (ca. 1650–1670) and detail (KIK-IRPA b67613). The detail shows a surveyor's instrument, similar to a Mordente compass, to the *left*; a Galucci type sector at the centre; a surveyor's circle to the *right*; and a Coignet surveyor's sector like instrument in the *top right*

The sector described in these manuscripts bears the same scales as the *reigle platte*. Judging by its appearance, it is a further development of the sector described in the Oxford and Paris manuscripts, although the sighting vanes have disappeared and we now have a much smaller calculating device, similar to Galileo's design.

There can be no doubt whatsoever that Coignet was one of the pioneers of sector-making. However, questions remain about the chronology of the tool's development. Coignet himself adds to the uncertainty by the fact that he refers to the different types of instruments as *pantometers*. To an extent, the chronology of development can be inferred from secondary evidence, but it is impossible to put forward exact dates with any degree of certainty.

Fig. 7.16 The use of a Coignet type compass for surveying purposes (Plates 31 and 32 in Marolois (1614–1615), EHC G 48942)

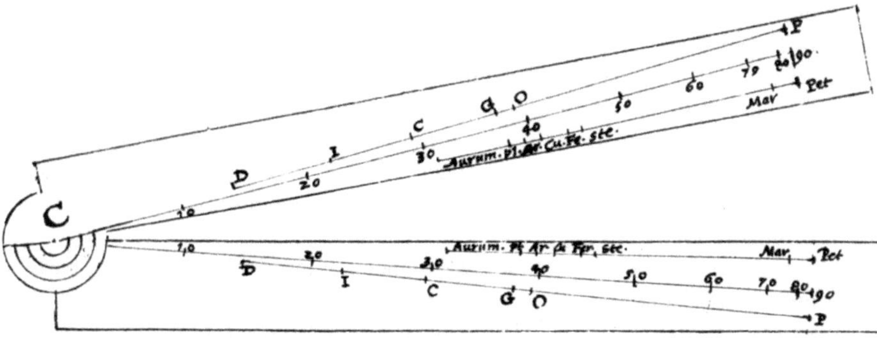

Fig. 7.17 Coignet (1618b)(EHC 520935)

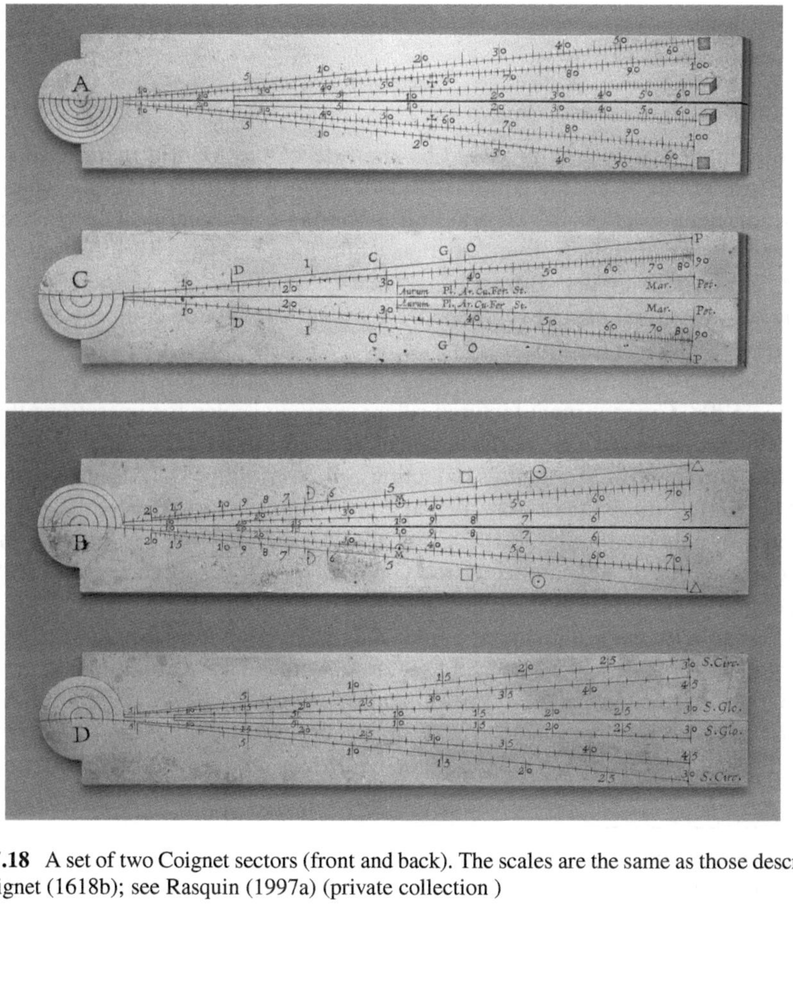

Fig. 7.18 A set of two Coignet sectors (front and back). The scales are the same as those described in Coignet (1618b); see Rasquin (1997a) (private collection)

Obviously the instruments themselves would potentially be an excellent source, but unfortunately just a few, undated, examples survive. Nonetheless, the resemblance of one of these instruments to the sectors described in the 1618 manuscript is so striking that there can be no doubt whatsoever that it originated in Coignet's workshop.[58] The Rijksmuseum in Amsterdam owns a sector supposedly signed by Gillis Coignet, Michiel's father, and dated 1562.[59] However, this is most likely a forgery, as this sector resembles so closely to the latter types of sector that it was presumably made in Coignet's (or his successor's) workshop after 1610. In a book on Galileo, D. Burger asserts that Coignet was producing sectors as early as 1584, and he mentions two sectors bearing Coignet's signature, one dated 1595, the other 1596.[60] It is not inconceivable, however, that Burger mistook proportional or reduction compasses for sectors, even if he appears to have actually seen the instruments. Unfortunately, he does not mention their location.

Hence, the only thing we know for certain is that, by 1610, Coignet had made some sectors, as is evident from a letter from Gloriosi, who writes:

> vetus quippe adinventum et ab omnibus una voce Michaele Coigneti Antverpiensi ut primo inventori attributum[61]
> *it is after all an old invention that is attributed unanimously to Michael* Coignet *from Antwerp*

Some indirect evidence is provided by the title of the Madrid manuscript. Here, Coignet refers to Mordente as a mathematician to Alexander Farnese, whereas in the Modena manuscripts he refers to him as coming from Salerno. This may suggest that the Madrid manuscript was written while Farnese was still alive, i.e. before 1592. This would support Burger's case that Coignet had already made sectors in the 1580s, albeit in adapted versions of Mordente's reduction compass. Other than that, though, the manuscripts offer few clues.

A tentative timeline for Coignet's instrument may be the following. In the 1580s, or late 1570s, Coignet manufactured rulers carrying scales for operations that are basically the same as those performed with a sector. In the late 1580s or 1590s, he became familiar with Mordente's reduction compass and/or Del Monte's sector.

Around the turn of the century, he integrated the two instruments into a surveying sector. Subsequently, in the 1610s, Coignet's sector shed its vanes, became smaller and quite similar in appearance to the Galilean sector. It still carried the scales of the rulers.

Alternatively, the scale carrying ruler may have been transformed into a sector prior to the 1600s. For the surveying sector, only those scales relating to surveying were transferred. Thus, a "new" instrument was developed, which could be presented to the sovereign.

Although both scenarios are possible, the former seems the more likely.

---

[58]Rasquin (1986–1988).

[59]ter Kuile (1986).

[60]Burger (1964, pp. 30–31).

[61]Galileï (1968, X, p. 363). Letter to Terrentius of 29 May 1610.

There is, however, an interesting story that at least points towards a varied parentage of the sector. It concerns Jan Eutel Zieckmesser (or Zugmesser), who was in the employ of Ernest of Bavaria, Prince-Bishop of Liège.[62] In 1603, during a stay at Padua, he claimed to be the inventor of a compass. It soon became clear that the features of his instrument resembled those of the Galilean sector. In fact, comparison revealed Zieckmesser's sector to be inferior to Galileo's (although the jury was of course biased). It is tempting to think of Zieckmesser's sector as an early Coignet-type instrument bearing only the scales of equal parts, sines and chords, as described in the Paris and Oxford manuscripts.

Among the users of Coignet's instruments were the Jesuits. Not only do we know that they purchased some instruments from his estate, but the file "mathematical instruments" of the Jesuit archive in the Rijksarchief actually contains some good drawings of Coignet-type sectors.[63]

In the late 1610s, the Jesuit college was an important centre for the study of mathematics (see p. 37). Its mathematics professor, Gregory of Saint Vincent s.j., was a pioneer of infinitesimal calculus. At least two of his students seem to have used Coignet-type compasses: Theodore Moretus s.j. and Joannes della Faille s.j. Theodore had joined the Jesuits in 1618 in Mechlin and studied at Louvain from 1621 onwards. In 1625, he became reader of mathematics at Munster, and subsequently he joined his teacher Gregory in Prague, where he resided from 1632 to 1639. In 1625, he helped Gregory write a mathematical treatise to convince the *Collegium Romanum* that his methods could be used to find a solution to the quadrature of the circle.[64] Towards the end of 1628, he left the Low Countries to assume several positions in Central Europe.[65] Possibly the "neunspitziger passer" manuscript came to Wrocław (Breslau) via Theodore Moretus, who at one time was stationed there.[66]

Joannes della Faille was sent to Spain, where he became a tutor to John of Austria the Younger, Philip IV's bastard son. During his time in Spain, Joannes would appear to have bought sectors from Jacob de Coster, who had previously worked in Coignet's workshop.[67] Among the papers of Joannes della Faille in the della Faille family archives, there is a manual on Coignet's sector. One of the manuscripts of Joannes's collection that remained in the library of the Jesuit order in Madrid was entitled *Fabrica y uso del Pantometro*, undoubtedly a manual on the use of Coignet's sector.

---

[62]Zieckmesser was referred to as "Fiamminghi", but this may denote anyone from the Low Countries (see Devisscher (ed.) 1995). On Ernest's court, see Halleux and Bernès (1995); on this particular story, see Schneider (1970, pp. 18–20).

[63]ARA AJ(VP), 1068.

[64]van Looy (1979, p. 5).

[65]van Looy (1979, pp. 17–18). On Moretus in Prague, see Bosmans (1928).

[66]Coignet (s.d.b).

[67]AFL 28.15.20, also AFL 28.15.31. Della Faille acted on several occasions as an intermediary for Spanish noblemen to obtain mathematical instruments in Antwerp. On della Faille, see Meskens (2005a).

Fig. 7.19 Next to della Faille's right hand, a Coignet-type sector can clearly be seen (Koninklijke Musea voor Schone Kunsten van België (Brussels), della Faille bequest, inv. no. 6254, photo KIK-IRPA B117502 (DYCK 6254))

Sir Anthony van Dijck's 1628/1629 portrait of Joannes della Faille features a Coignet-type sector, making it the sector with the largest viewing audience (Fig. 7.19).[68]

As late as 1638, Englishman Ignace Stafford (1599–1642), who taught at the Royal Academy in Lisbon, wrote a treatise on *Arithmetica practica geometrica logarithmica* in which he discusses the use of a *pantometer*, as he calls the sector.[69] This indicates that Iberian Jesuit mathematicians came into contact with the sector as a calculating instrument through Coignet's work. This may have involved Joannes della Faille or any other Flemish Jesuit who came to Portugal in the hope of joining the missions to China. One such Jesuit was della Faille's fellow student with Gregory of Saint Vincent, Jan Ciermans s.j., who acted as a military adviser to

[68]See Meskens (1999).

[69]See Gessner (2011).

the Portuguese during their war of independence with Spain.[70] As for the parentage of the sector, Stafford claims [71]

> *The first inventor of the sector, such claim the French, was Michiel Coignet; the Italians want it to be Galileo. The first of these two, who achieved the renown of a good Geometer, could accomplish and change it in a way so it looked like a new work and invention, the latter deserves the infamy of great thief of others instruments. And so neither the one nor the other will retain the honour that is due to the first inventor, because Father Clavius – much earlier than both of them – shows at the beginning of his* Geometria practica *the design of an instrument which he calls 'instrument of parts', where there are sketched out the first shapes of the sector, and over this foundation Coignet could very easily raise his building.*

This appreciation by Stafford may not be too far from the truth, if one includes among the earliest inventors the likes of Jost Bürgi, Paolo Galucci, Guidobaldo del Monte, Jacques Besson, …

[70]Jan Ciermans s.j. (also known as Joao Pascásio Cosmander) had been a student of Gregory of Saint Vincent s.j. in 1623–1624. He had entered the Society in 1619. Ciermans had come to Portugal in 1641 en route to the China mission. Having heard about his engineering talent, the Crown assigned him to supervise the construction or rebuilding of the Kingdom's fortifications. He participated in several battles. Father-general Vitelleschi s.j. ordered him not to participate in any operation that would compromise his neutrality. The King, however, appointed him to the rank of colonel and chief engineer, which led to his expulsion from the Society. In 1547, he was wounded and captured by the Spanish at Elvas. A year later, he was forced to join the Spanish Army which Philip IV had ordered to attack Olivença. There, Ciermans was mortally wounded and died on 20 June 1648. Alden (1996, p. 106), van Looy (1979, pp. 19–20), Van de Vyver (1975), Dhombres and Radelet-de Grave (2008).

[71]Translation from the Portuguese in Gessner (2011, p. 223).

Chapter 8
The Art of Navigation

8.1 Introduction

In 1579, Coignet wrote the book for which he is still best known today: *Nieuwe Onderwijsinghe op de principaelste puncten der Zeeuaert* (New Instruction on the principal points of navigation). It was published by Hendrik Hendriksen[1] in 1580 as an appendix to Merten Everaert's translation of *Arte de Navegar* by Pedro de Medina. The French version, *Instruction nouvelle*, was published the following year as a book in its own right. Apart from Waghenaer's naval atlases, these were the only books on navigation to be published in Antwerp or the Southern Netherlands. Medina's and Coignet's books would subsequently be reprinted on three occasions by Cornelis Claesz. from Amsterdam (1589, 1592, 1598).

The book is dedicated to Gillis Hooftman, one of the wealthiest merchants and shipowners in the Low Countries (Fig. 8.1). He was probably of Rhenish descent[2] and, after 1566, fell under suspicion of Protestantism.[3] Hooftman's ships traversed all of Europe's waters. Together with Gerard Gramaye, he founded a company that traded with the Baltic, and, from 1577, he was also involved in navigation to Northern Russia and Africa.[4] His circle of friends included William of Orange and

[1]Hendricksen ran a printing shop in Antwerp from 1572 to 1588 and, for a short while in 1575–1576, also in Louvain. Some 14 books printed at his shop have thus far been identified, including a number of prognostications. However, from his transactions with Plantin, we know of around 30 titles. In the 1585 lists of the Civil Guard, he is categorized as a Protestant (SAA A4830(1), f26*r*). The following year, he was tried for having printed "seditious books". Rouzet (1975, p. 91).

[2]Prims (1932).

[3]Prims (1932) and Wegg (1979, p. 174). His son, who was an alderman on the City Council under Philips of Marnix, left Antwerp in 1584 and moved to Delft (van Roey 1985a). Hooftman's daughter attended Heyns's school.

[4]Denucé 1937, p. 14.

A. Meskens, *Practical Mathematics in a Commercial Metropolis: Mathematical Life in Late 16th Century Antwerp*, Archimedes 31, DOI 10.1007/978-94-007-5721-9_8,

© Springer Science+Business Media Dordrecht 2013

Fig. 8.1 Maerten de Vos, *Portrait of the shipowner Gillis* Hooftman *and his wife Margaretha van Nispen*, July 1570 (Rijksmuseum (Amsterdam), SK-A-1717)

Thomas Gresham (1519–1579).[5] Hooftman's collection of maps provided the basis for Ortelius's *Theatrum Orbis Terrarum* (1570), so it comes as no surprise that he also sponsored Coignet's work.[6] He died at the beginning of 1581.

8.2 The Art of Navigation

Up until the sixteenth century, Flemish navigation was largely restricted to a small sailing area, requiring relatively little nautical knowledge. Ships never ventured far from land, and sailors would simply learn by experience. Among the limited skills required were the ability to recognize features of the coastline and to distinguish between currents.

If a mariner was literate, he might have possessed a Rutter, a book providing information about coastal profiles, the tides, and the position of sandbanks alongside sailing directions. Many such Rutters were published in the Netherlands (Fig. 8.2).[7]

[5] Denucé (1937, p. 14) and Van Acker (1982, p. 80).

[6] Meskens (1992, 1994b, pp. 126–127).

[7] Mörzer-Bruyns (1984).

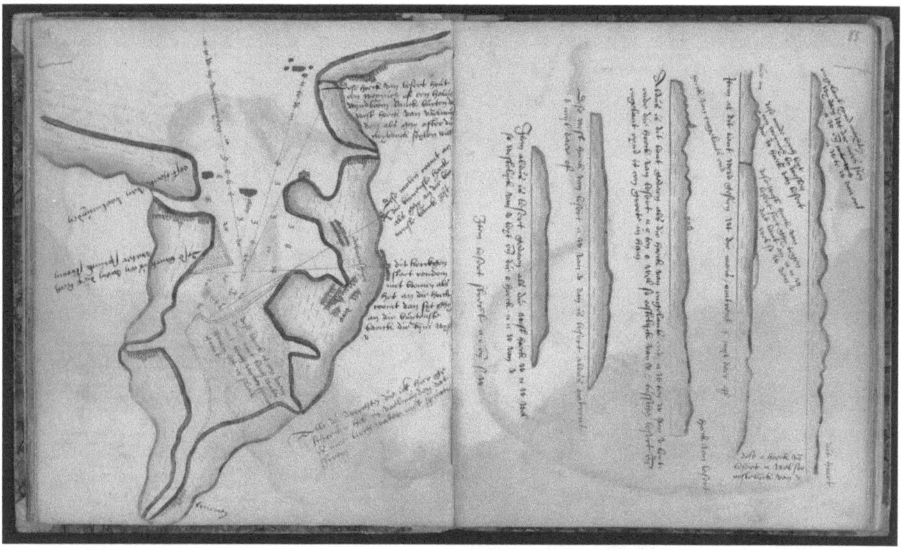

Fig. 8.2 Two pages from the sixteenth-century manuscript *Zeeboek* (EHC B 29166), a map of Falmouth Bay (*left*) and coastal profiles of the vicinity of Lizard (*right*) (see Denucé and Gernez 1936, pp. 19–22)

A shipmaster may also have relied on a Portolan chart, a navigational map of a particular area (Fig. 8.3). Portolan charts featured parallel circles of latitude as well as meridians as equidistant straight lines (see also Chap. 9), implying that, over larger distances, the picture provided was somewhat distorted. A course along a constant compass bearing drawn as a straight line would diverge increasingly from the actual course as the distance grew. This was due to the fact that a Portolan chart is not conformal, i.e. the projection preserves neither angles nor areas. From this it follows that a loxodrome, which intersects each meridian at the same angle, does not project onto a straight line. For a long time it was thought that a loxodrome was a large circle, i.e. that keeping the same bearing all the time would eventually bring one back to one's port of departure (assuming, of course, that no land intervened). However, it was noted by the Portuguese scholar Pedro Nuñez (1492–1578) in 1537 that a loxodrome is in fact a spherical spiral with the poles as asymptotic points.[8]

In the first Dutch navigation manual, *Onderwijsinghe vander zee*, which was published at Amsterdam (presumably in 1544, though only the third impression from 1558 is known[9]), Cornelis Anthonisz. deals with the compass, the cross-staff and the quadrant.[10] With the compass, one could check one's course; cross-staff and quadrant were for measuring the altitude of a celestial object above the horizon

[8]Waters (1978, pp. 71–72) and Leitão (2007).

[9]Davids (1986, p. 73).

[10]Keuning (1950).

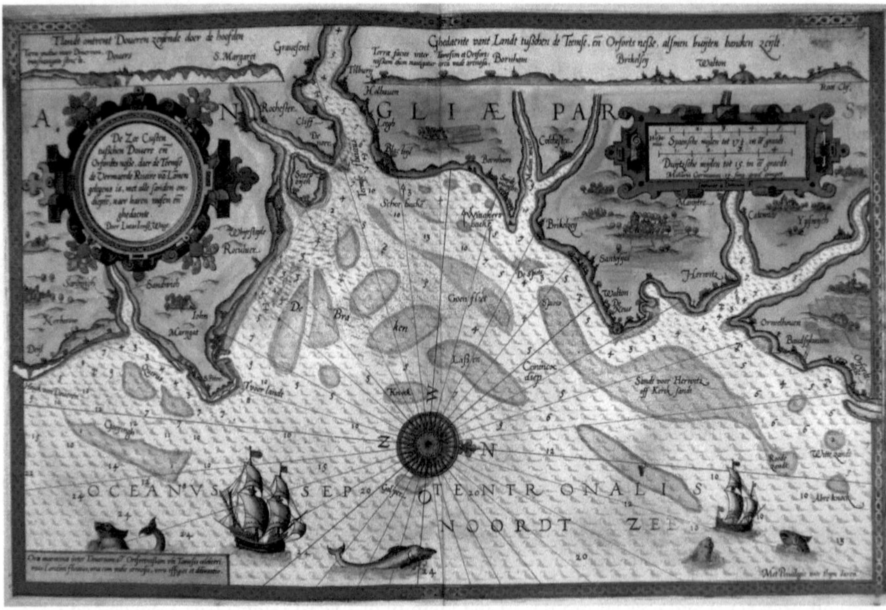

Fig. 8.3 Lucas Jansz. Waghenaer, *The English Channel and its coasts between Nieuwpoort and Dieppe* (1584). Notice the silhouettes of the coast (From Wagenaer 1584, MPM R 45.2)

(in most cases the sun or the North Star) with a view to determining latitude. With a simple rule known as the "Regiment of the North Star", one could convert the altitude of the North Star into true latitude.

Determination of longitude posed more of a problem: there was no simple astronomical rule, because, owing to the Earth's rotation a fixed point of reference was lacking. However, soon after the problem became acute (due to navigation to the Indies and the Americas), it dawned that the essence of the problem was timekeeping.

Gemma Frisius[11] proposed the following theoretical solution: suppose a vessel has sailed from harbour *A* and has reached a point *B*. The local time at *B* can be determined by using the position of celestial objects. If one knows the time at *A*, at the same instant, the time difference between *A* and *B* indicates the difference in longitude. Indeed 1° longitude difference equals 4 min. The method could not be put into practice, however, until the construction by John Harrison in 1761 of the first nautical chronometer.[12]

Another solution described by Frisius (Fig. 8.5 right), but previously proposed by Apianus (Fig. 8.5 left), was the lunar distance method (Fig. 8.4). This method uses

---

[11]Pogo (1935).

[12]Sobel and Andrewes (1998) and Kennerley and Seymour (2000).

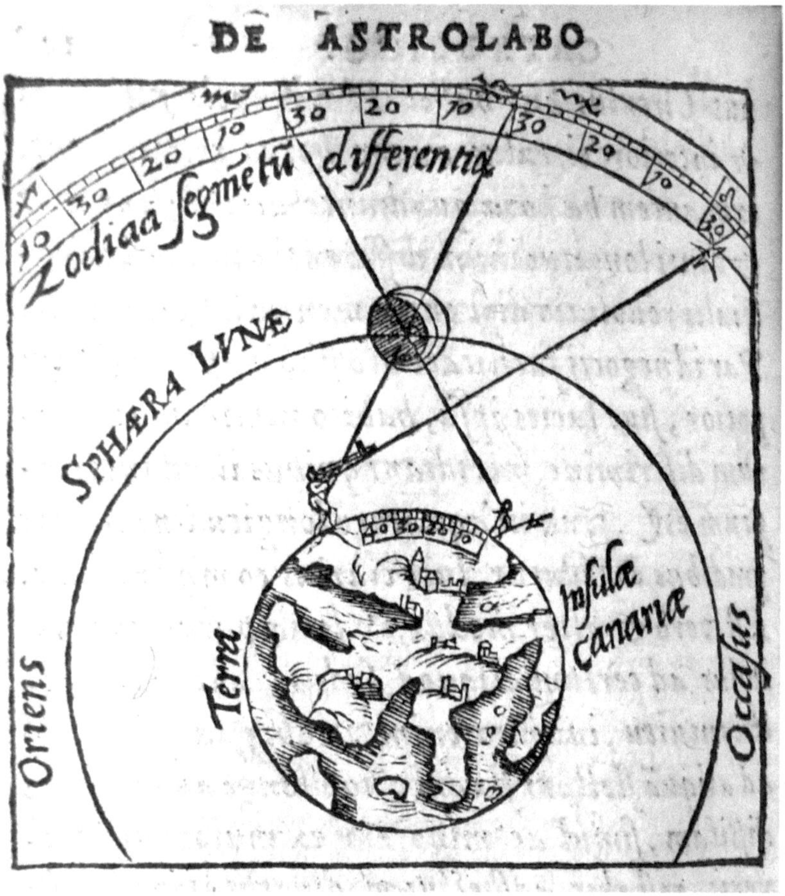

Fig. 8.4 Determination of longitude using the lunar distance method in Frisius and Apianus (1533) (MPM A 2794.2)

the relatively fast movement of the Moon along the celestial sphere. In 27.3 days, the Moon completes one orbit of the Earth. Hence, in 1 h, it moves half a degree of arc. By measuring the angle between the moon and another celestial body (preferably a fixed star) and by comparing this value with tabulated values for a fixed position (e.g. the harbour of departure), a navigator can determine not only his local time but also the time in this fixed position. Unfortunately, though, the inaccuracy of existing instruments meant that the observations made were of no use.

In other words, a navigator on the high seas was unable to calculate either his position or the distance previously covered. Therefore, he had to rely on guestimates (a process known as "dead reckoning"). One technique for sailing the Atlantic was to follow a parallel of latitude until land was sighted and to subsequently make for the port of destination.

Fig. 8.5 Petrus Apianus (l.) and Gemma Frisius (r.) (MPM 4–97)

The age of voyages of discovery demanded an improved nautical ability. As the Spanish and the Portuguese were the first to venture far into the ocean, it is not surprising that they produced the first pioneers of navigation, such as Pedro de Medina, Martin Cortês, Pedro Nuñez and Rodrigo Zamorano.

By the third quarter of the sixteenth century, the farthest the Dutch and the Flemish had ventured was the Canaries to the south and Norway and the Baltic to the north, as indicated by the tide table in Coignet's *Nieuwe Onderwijsinghe*. Nonetheless, as early as 1560, a few enterprising merchants had begun to explore routes to Russia and the White Sea.[13] After the lifting of the blockade of the Scheldt by the Sea Beggars in 1576, some even embarked on voyages to Africa.[14]

It is easier to explain why no navigation manuals were published at Antwerp after 1585. After the city's surrender to the Spanish, the Scheldt was once again blockaded, depriving Antwerp from its access to the open sea for generations to come. Hence, demand for navigation manuals simply dissipated. Moreover, many of the printers who possessed the expert knowledge to produce such manuals had by then fled to the north. It is no surprise, then, that the three subsequent reprints of Coignet's book were published in Amsterdam, which by that time had become the principal port of the Low Countries.

---

[13]De Smet (1975–1976, pp. 18–19), Wijnroks (2003), passim.

[14]van der Wee (1987, p. 14) and Denucé (1937).

8.3 De Medina's *Arte de Navegar*

Pedro de Medina was an important author on the science of navigation.[15] Born in 1493 to parents in the service of the Duke of Medina Sidonia, he would himself come to serve the Duke for a while, as a librarian.

De Medina wrote several books in his 74-year life, including the pioneering *Libro de grandezas y cosa memorables de España* (1548), which in today's terms might be referred to as a tourist guide, and a book entitled *Dialogo de Verdad* (1555). The latter would be reprinted no fewer than 13 times in the course of the sixteenth century. For his employer, he wrote a history of the Medina Sidonia family (*Crónica de los duques de Medina Sidonia* (1561)). His manuscript *Libro de Cosmographia* saw the light of day in 1538.[16]

De Medina earned his living as an instrument maker and cartographer. In 1539, he was commissioned to make maps and instruments for the *Consejo de Indias* and the *Casa de Contratación*, and he was appointed as an examinator of navigators and seamen. His relationship with the *piloto mayor* or "master navigator" of the *Consejo*, the explorer Sebastian Cabot, was troubled and actually ended in a lawsuit.

De Medina first published his groundbreaking work on compass navigation, *Arte de Navegar*, in 1545 (Fig. 8.6). This was followed in 1552 by an abridged version, entitled *Regimiento de navegación,* which was subsequently reprinted in 1563.

Arte de Navegar discusses the use of a navigator's basic instruments, notably the mariner's astrolabe, the compass and the cross-staff. It also explains the use of maps and the measurement of the altitude of the sun and the North Star, for the purpose of determining latitude (see Fig. 8.9). De Medina stressed the importance of navigation to extending the Spanish dominion. He argued that it was necessary to provide those who would be crossing the ocean or charting unknown territory with proper instructions. The book almost instantly became the standard training manual for navigators, not just in Spain, but across Europe. It was translated into French, Italian, German, English and Dutch, and appeared in at least 27 editions. Francis Drake is known to have possessed a French copy, and Willem Barentz left behind a Dutch copy on Novaya Zemlya.

The Dutch version was a translation by Merten Everaert (Bruges 1540 – Leiden or Delft, after 1600). Everaert was a prolific translator. Among the works he rendered in Dutch were writings of Paracelsus and the Swiss physician Ruffen on medicine, a work by Justus Lipsius, a Portuguese book on the Kingdom of Congo, parts of Plutarch, and Zamorano's navigational book. He also translated Waghenaer's *Spieghel der Zeevaerdt* into Latin. In the 1560s, he worked as a translator and copyist at Plantin's print shop, i.e. he reproduced manuscripts in a more legible handwriting for the benefit of the typesetters.[17] The only surviving work in his hand

[15] See Crone (1953), Waters (1978), passim, Verlinden (1987).

[16] Lamb (1972).

[17] See Voet (1980–1983, nos. 1081, 1097, 1140, 1579, 2091, 2183, 2416, Add. 606, 1081, 1097, 2183).

Fig. 8.6 Detail of the frontispiece of de Medina, *De Zeeuaert,* 1580 (EHC G 16623)

is *Ephemeridae meteorologica.* Everaert eventually emigrated to Delft, where he was said to have been "a monk, born in St-Omer in the land of Walloons".[18]

For his Dutch version of *Arte de Navegar*, he not only translated directly from the Spanish, but also made use of the French translation (1554) by Nicolas Nicolaï[19] (1517–1583).

8.4 Nieuwe Onderwijsinghe

Coignet's *Nieuwe Onderwijsinghe op de principaelste Puncten der Zeevaert* was published in 1580 as an appendix to De Medina's *De zeevaert, oft: Conste van ter zee te varen*, published by Hendrik Hendriksen. A year later, Hendriksen published the extended French translation, *Instruction nouvelle des poincts plus excellents & necessaires, touchant l'art de naviger*, as a separate book (Fig. 8.7).

In his preface, Coignet is very commending of Gillis Hooftman, suggesting that he wrote or published the book at the behest of the latter. Looking beyond

[18]Briels (1971, p. 123).

[19]A French soldier and geographer who was appointed as Royal Geographer by Henry II. He is the author of several "Descriptions" of French regions and a navigational work.

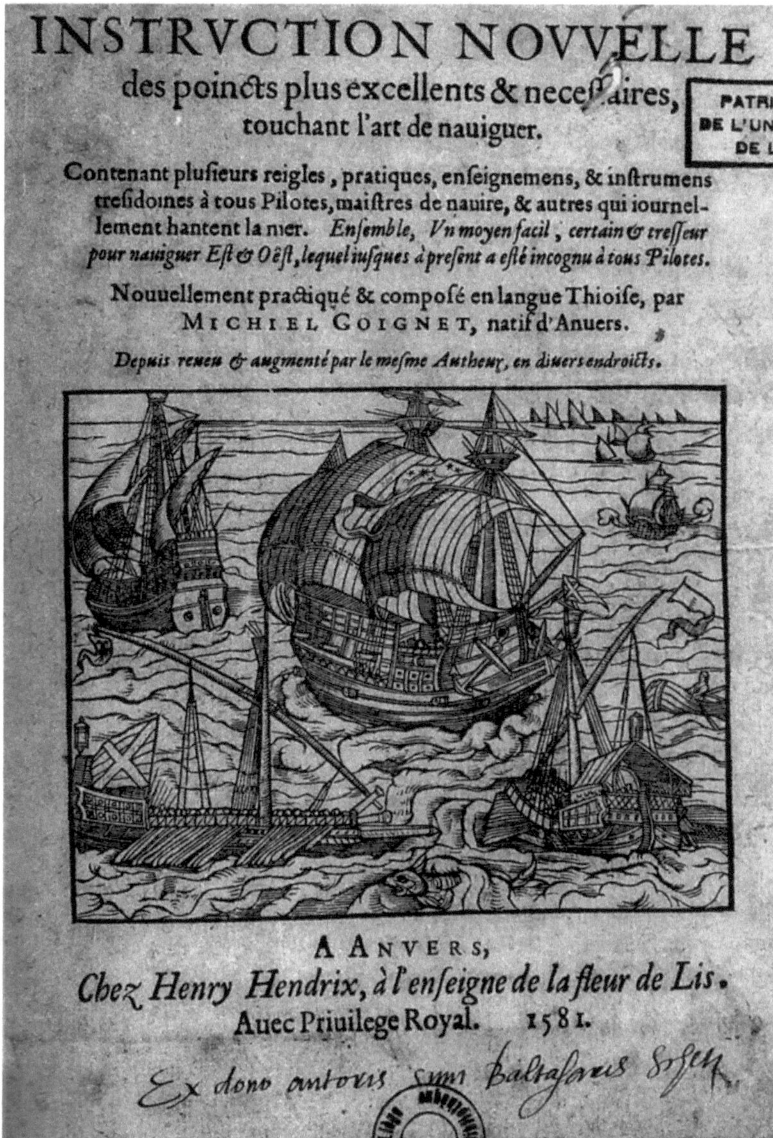

INSTRVCTION NOVVELLE
des poincts plus excellents & necessaires,
touchant l'art de nauiguer.

Contenant plusieurs reigles, pratiques, enseignemens, & instrumens tresidoines à tous Pilotes, maistres de nauire, & autres qui iournellement hantent la mer. *Ensemble, Vn moyen facil, certain & tresseur pour nauiguer Est & Oëst, lequel iusques à present a esté incognu à tous Pilotes.*

Nouuellement practiqué & composé en langue Thioise, par
MICHIEL COIGNET, natif d'Anuers.

Depuis reueu & augmenté par le mesme Autheur, en diuers endroicts.

A ANVERS,
Chez Henry Hendrix, à l'enseigne de la fleur de Lis.
Auec Priuilege Royal. 1581.

Ex dono autoris sum Baltasaris Schetz

Fig. 8.7 The copy of *Instruction nouvelle* (Université de Liège, BGPhL-CICB, R311B) given by Michiel Coignet to Balthazar Schetz, the maecenas of his 1573 book *Cent Questions Ingénieuses* (see p. 59, footnote 5)

the rhetoric, the preface reveals that things were already changing in the port of Antwerp. The city's outlook had begun to broaden as the port was preparing for expansion, with ships now also sailing to the Americas and the Indies. However, as political events unfolded, Antwerp would be unable to deliver on its promise.

The book may be considered to consist of three parts. The first four chapters deal with general seafaring knowledge: the basic principles of cosmography, the winds, the compass and its variation, and the charts. This part is mostly theoretical. In the subsequent chapters (5–16), he describes some navigational instruments and how to use them. He provides a description of the mariner's astrolabe, the nautical hemisphere, the cross-staff and the nocturnal. The final chapters deal with position finding (17, 18, 20) and the tides (19).

Few of the instruments Coignet describes were new. Those that were may be referred to as analogue calculating devices, intended to replace calculations based on tables with a mechanical procedure that instantly yields the desired information. Such aids need not have been complicated contraptions; some essentially consisted in slight alterations or modifications to previously existing instruments.

One such alteration is an arc he added to the alidade of the mariner's astrolabe. The mariner's astrolabe[20] was a late fifteenth century Portuguese development. It was especially designed for measuring the altitude of the sun. In comparison with an astrolabe, a mariner's astrolabe is a very crude instrument. It is essentially a disk (usually made of brass) with a revolving ruler or alidade. At the ends of the alidade, there are tiny pin-hole sighting vanes. To measure the altitude of the sun, one turns the alidade into such a position that the rays of the sun pass through the pin-holes in both vanes (Fig. 8.9). At the end of the alidade one can then read the altitude (or its complement) on the graduated circumference. To reduce the effects of wind on the disk, the brass was pierced or fretted (usually one hole in each quadrant). On the other hand, the bottom was made heavier, in order to increase the mariner's astrolabe's stability and enhance its use.

Coignet's mariner's astrolabe seems to have had all these features and an additional one that allowed users to immediately read the latitude. At the ends of the alidade, one notices little arcs, with an engraved scale: 0° at the centre and 23.5° at the ends. This allows one to take immediate account of the solar declination: latitude is found in either the upper or the lower parts of the arc, depending on whether it is winter or summer (see Fig. 8.8).

There are no extant mariner's astrolabes with such arcs. However, a similar mariner's astrolabe was described by Jacques Devaulx in his *Premières Oeuvres* in 1583.[21] One may speculate whether or not Devaulx knew Coignet's book and copied this part or whether this type of astrolabe had already come into use in navigation during the 1580s.

---

[20]On the mariner's astrolabe see: Anderson (1972) and Stimson (1988).

[21]Stimson (1988, p. 36).

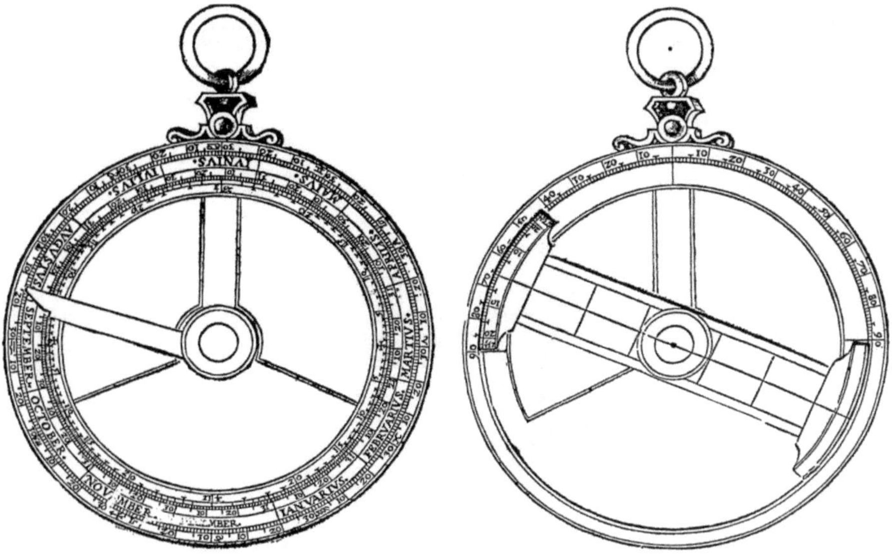

Fig. 8.8 Coignet's mariner's astrolabe, back (*left*) and front (*right*) (EHC G 16623). To make an observation with Coignet's astrolabe, one first has to calculate the solar declination. The back of the astrolabe is a computing device for precisely that purpose. It is engraved with four concentric circles. The outermost circle is divided into 12 × 30°, each section of 30° representing a sign of the zodiac. The second circle is divided into 12 months, the third into days. On the fourth circle are engraved the values for the solar declination of each day. To use the instrument, one puts the ruler at the day of the year and reads the corresponding declination on the fourth circle.

Using the front, one can now measure the Sun's altitude. The scale on the rim indicates the complement of altitude (i.e. it has 0° at the top) thus, at the solstices, giving an immediate reading of latitude. With the aid of the little arcs at the end of the alidade, one can determine latitude at other times. The arcs bear a scale of degrees up to 23.5°. Using the calculated declination, one can read one's latitude next to the declination on the arcs (the upper part for winter, the lower part for summer)

Coignet's is the first book in which we find the description of a cross-staff with more than one transom, which made it possible to measure smaller angles.[22]

The cross-staff is an instrument whereby one can measure the altitude of a celestial object, more often than not the North Star. It seems to have been introduced into nautical practice by the Portuguese around 1515.[23] The device had previously been used as an astronomical aid, known as a *baculus jacob*, as early as the fourteenth century. The cross-staff consists of a graduated wooden staff along which slides the transom. To make an observation, one needs to put one end of the staff to one's eye, in a manner such that the transom is at a vertical angle with the horizon. The staff should, moreover, always be held so that the lower end of the transom is

[22]Waters (1978, pp. 204–205). On Coignet's cross-staff, see Meskens (1992, 1994b, pp. 129–130) and Mörzer-Bruyns (1994, pp. 28–29).

[23]Mörzer-Bruyns (1994, p. 14).

Fig. 8.9 The use of the mariner's astrolabe in de Medina (1580) (EHC G 16623)

aligned with the horizon. The transom is then moved along the staff until its upper end is aligned with the object, a process whereby the staff is rotated slightly, while the lower end of the transom is kept aligned with the horizon. The eye thus moves up and down continuously. Once the transom is in the right position, the altitude can be read from the intersection of the transom and the staff.

In his book, Coignet proposes three transoms (see Fig. 8.10). With these angles, one measures between respectively 90° and 30°, 30° and 10° and less than 10°. In the French translation, he also provides a description of Frisius's cross-staff, which was designed to measure small angles. Frisius had originally intended this cross-staff to be used in measuring the distance between stars. Coignet's idea of adapting Frisius's staff for navigational purposes was not new though. William Bourne had already made such modifications in 1574.[24] The suggested use was most probably inspired by the requirements of northern navigation. Perhaps Hooftman's trade with Russia provided a reason for writing about it.

A nocturnal (see Figs. 8.13 and 9.2) is an instrument that allows one to tell the time during the hours of darkness. They were made of brass or wood, the latter being preferred by seamen.

[24] Waters (1978, p. 135).

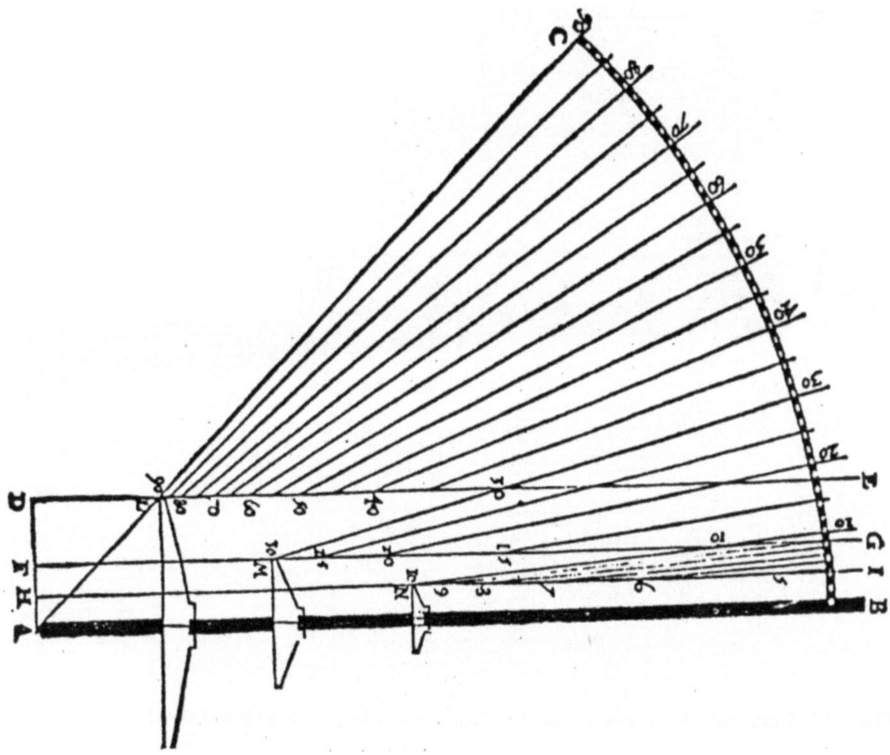

Fig. 8.10 Geometrical construction of the graduation on Coignet's cross-staff (Coignet 1580, EHC G 16623). In practice, only one transom slides along the staff

The device essentially consists of a disk fitted with a handle, a smaller disk with a pointer on top of the first, and a third, moveable disk with an extended arm. These disks are pierced with a nut and bolt to hold them together. The outer disk is divided into months and days and/or into the zodiacal signs. To make an observation, the pointer of the middle disk, which is divided into hours, is set at the observation date on the outer disk. The instrument is held at arm's length and turned towards the North Star, which is observed through the central hole. The extended arm is now turned to align with either Kochab (βUMi) in Ursa Minor or Dubhe and Merak (αUMa and βUMa) in Ursa Major (also referred to as the "Guards" in sixteenth-century books). Time can now be read from the position of the arm on the second disk. The teeth cut at the hour position make it easier to read the time in darkness (see Figs. 8.13 and 9.2).

Derived from the meteoroscope, the nautical hemisphere is an instrument that appears to have been introduced to navigational practice by Coignet himself.

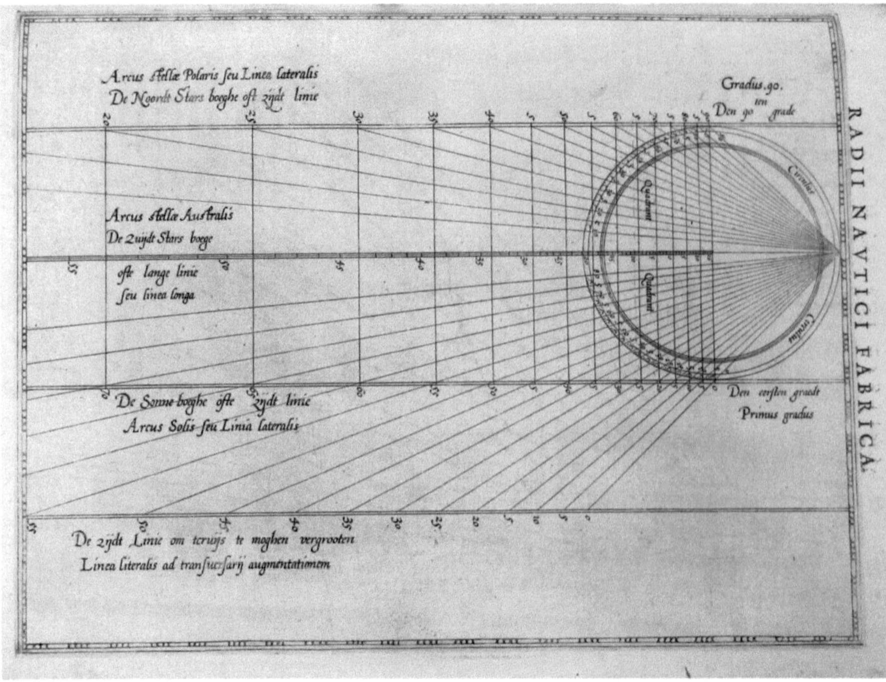

Fig. 8.11 Geometrical construction of the graduation in Wagenaer (1584) (MPM A549)

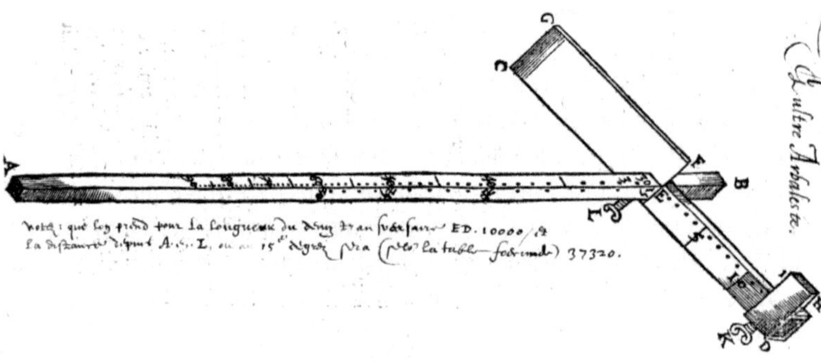

Fig. 8.12 Frisius's cross-staff for use at high latitudes (Coignet 1581, EHC G 16623)

A meteoroscope is a device that allows one, among other things, to determine the latitude of a position irrespective of the time of day. Little is known about the instrument other than what Pappus and Regiomontanus mention about it in their commentaries on the writings of Ptolemy. Coignet, too, knew about the

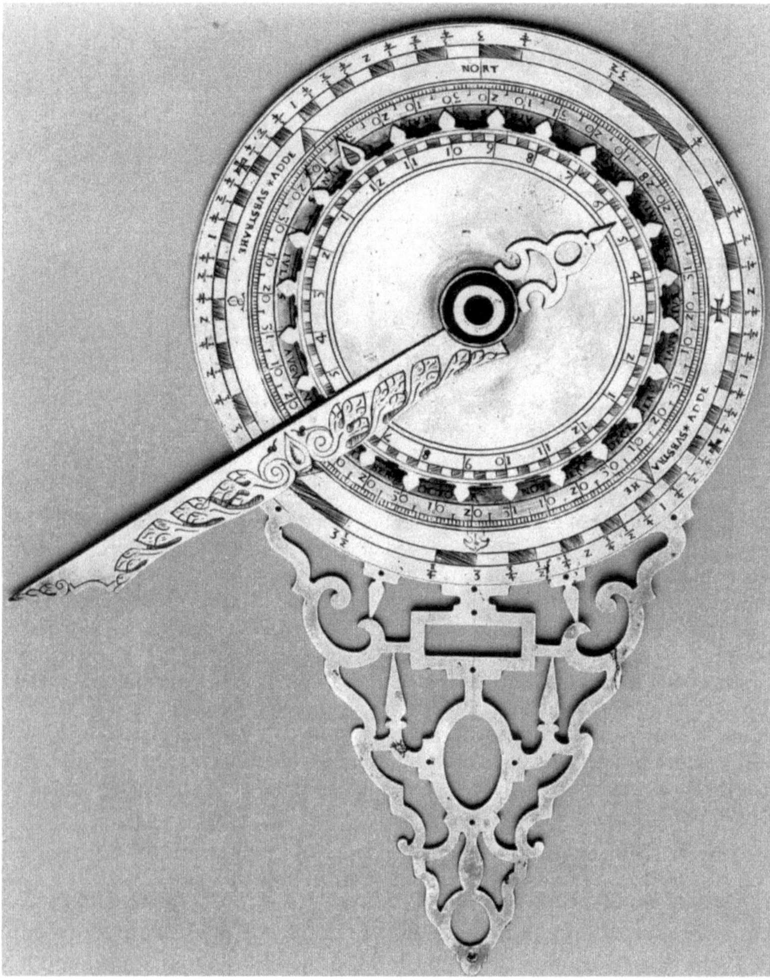

Fig. 8.13 Nocturnal, front (private collection)

instrument through the work of Regiomontanus. The latter had described it in a 1462 manuscript that was presented to Cardinal Bessarion.[25] In 1514, Johannes Werner included it in his *Nova translatio libri geographiae Cl. Ptolomaei*.[26]

---

[25]Zinner (1968). During the Italian Renaissance, Cardinal Basileios Bessarion (1403–1472) was the embodiment of the merger between East and West. In 1437, he was appointed Metropolitan of Nicaea, in which capacity he attended the Council of Florence. From 1440, he lived in Italy. His library would form the nucleus of the Biblioteca Marciana. On a visit to Vienna in May 1460, he invited Peurbach and his student Regiomontanus to travel back to Italy with him. Regiomontanus joined the Cardinal's retinue for the following years.

[26]Zinner (1968, p. 314). For a reconstruction, see Rome (1927, p. 91.)

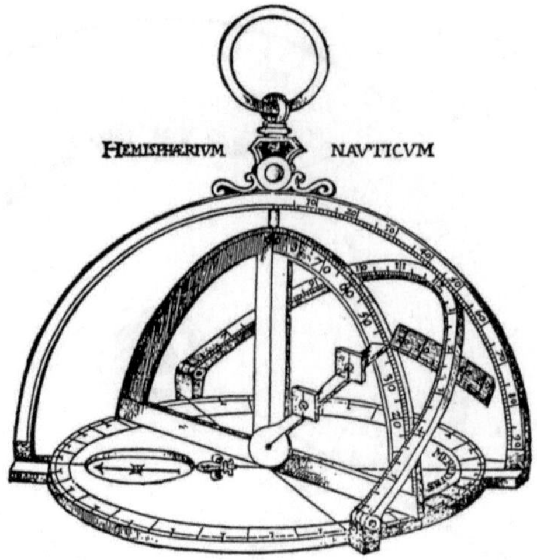

Fig. 8.14 Coignet's nautical hemisphere. Assuming that the observation was made in the northern hemisphere, the drawing suggests this happened at 48° N. The declination is 23.5°, indicating it was made in June at about 2 h 30 min.

In order for a nautical hemisphere to be useable, it needed to be suspended so that it could hang freely. Presumably one man had to hold the sphere while another performed the measurements. With the aid of a compass, the NS line is oriented towards the north. Subsequently, the altitude arc is turned so that its shadow forms a straight line. The alidade is turned in such a way that the sunrays pass through both pinholes. The little arc representing the declination is moved towards the alhidade. The arc representing time is then turned in such a way that the alhidade points to the known declination. At the intersection of the time arc or equinoctial arc and the altidude arc, one then reads the time, while at the intersection with the meridian arc one reads the latitude (Coignet 1580, EHC G 16623)

The nautical hemisphere (Fig. 8.14) consists of a round horizontal plate with a compass inset. Perpendicular to the plate, along the NS line, an arc representing the meridian is erected. Another arc, representing the equinoctial, can revolve around the EW axis. Onto this arc is engraved an indication of time, between 6 a.m. and 6 p.m. Perpendicular to and on the inside of this arc, there is a small moveable second arc, representing the solar declinations. A third arc, representing altitude, is also perpendicular to the plate, but can revolve around the pole of the sphere. On this arc, using the alidade, one can read the altitude of the sun (in this instance, one measures the complement of the altitude).

In *Instruction nouvelle*, Coignet adds a numerical example to the chapter on the nautical hemisphere. He gives the observations for 11 May (sun's right ascension and declination) and calculates, with the aid of spherical triangles, the latitude and the time. He then describes the required operations with the nautical hemisphere and arrives at the same results.

The nautical hemisphere already seems to have been in use in the 1580s. The Séminaire of Tournai owns a Coignet-type nautical hemisphere, though most of

Fig. 8.15 The remains of a Coignet-type nautical hemisphere, presumably from Coignet's own workshop (Bibliothèque et Archives du Séminaire de Tournai). See Lefrancq (2008)

the arcs are unfortunately missing (Fig. 8.14).[27] William Barlow (1597)[28] and Edward Wright (1599)[29] developed the instrument further. There are indications that instruments of this type were used on ships of the Dutch East India Company.[30] The *Casa de Contrataccíón* at one time owned a nautical hemisphere by Coignet.[31] Nautical hemispheres were listed in the inventory of seaman's instruments intended for use on Van Neck's second expedition to the Indies.[32] Fierce criticism came from William Borough, because Coignet had not accounted for the compass variation that he had described earlier in his book.[33]

To determine longitude, Coignet proposed to use Frisius's method in conjunction with hourglasses suspended in gimbals, which had to be turned around daily at noon of the place of departure. When the hourglass is turned, one records the time at the position reached (by observing the sun). From the time difference, one can then

27The only known complete surviving hemisphere was made by Charles Whitwell and is kept in Florence at Museo della Storia della Scienza.

28In 1597, William Barlow published *The Navigator's Supply*, in which he describes a nautical hemisphere resembling Coignet's hemisphere to some extent. Barlow (1597).

29Waters (1978, p. 316). Edward Wright (1561(?)–1615) was an English mathematician and cartographer and author of the book *Certaine Errors in Navigation* (1599), in which the mathematical basis of the Mercator projection was explained.

30Davids (1986, p. 425).

31Inventory of Pulido Rubio, see Destombes (1969, p. 5).

32Davids (1986, p. 425). Jacob van Neck (1564–1638), together with Wybrand van Warijck and Jacob van Heemskerck, led the second Dutch expedition to "East India" (1598–1599). See Terpstra (1950) and van Foreest (1980–1981).

33Borough (1581), f Gjr ff.

calculate longitude. Although the method is theoretically correct, hourglasses can hardly have yielded satisfactory results.[34]

He also proposes a theory whereby (if correct) the longitude can be determined on the basis of magnetic variation. By the fifteenth century, the realization had grown that a compass needle did not point north but east of the north (a phenomenon referred to as easting).[35] Sailors in the Mediterranean did not take this into account. They simply aligned their needle along the NS axis. In the Low Countries, however, compasses were being adjusted in such a way that when the needle was pointing to the magnetic north the compass card was pointing to the true north.

Fig. 8.16 Mercator (Mercator and Hondius (ill) (ca.1607), MPM 8 320)

The fact that magnetic variation differed from place to place raised hopes that it could solve the problem of longitude. Martin Cortês was one of the first to propose a theory to explain variation. He postulated a magnetic pole near but not coincident with the true North Pole. The phenomenon of variation was described in *Cosmographicus* (Landshut, 1524) by Petrus Apianus, who mentions an easting of 8° E.[36] According to Coignet's observations, easting amounted to 9° E in Antwerp.[37] Coignet's theory goes back to Mercator (Fig. 8.16), who mentions it on his 1569 world map.[38] The magnetic declination was said to be 0° at the Azores, more specifically on the island of Corvo. By adding his own observations, Mercator concluded that the magnetic north was 16.5° "behind the pole".[39] Coignet asserted that if Mercator's

---

[34]Determining local time on the basis of the sun's position can result in errors of up to 15′ (Waters 1978, p. 60). In order to be able to determine longitude to within half a degree, the error should not exceed 2 min! Furthermore, there is a degree of uncertainty in determining time by means of an hourglass.

[35]Ironically de Medina (1580), in his third chapter, rejected magnetic variation.

[36]Röttel (1996, p. 147).

[37]Davids (1986, p. 59), mentions an Easting of 6°.

[38]Cartouche at the top of the map at about 160° E.

[39]I.e. the pole is on the meridian of the Azores. The Azores, North Pole and magnetic pole are situated in that order, and the arc between the North Pole and the magnetic pole is 16.5°, thus the latter is situated at 73.5° N and (using present-day values) 149° E. In a letter from 1547 to Perrenot de Granvelle, Mercator elaborates the first of several attempts to determine the exact coordinates of the Arctic magnetic pole. He used two observed variations, Danzig (14° northeasting) and Walcheren (9° northeasting) to orient two great circles, one through each of these places. The intersection of these great circles would give the coordinates of the magnetic pole. Later, adopting 16°44 northeasting at Regensburg, and using respectively the meridian through the Azores and the

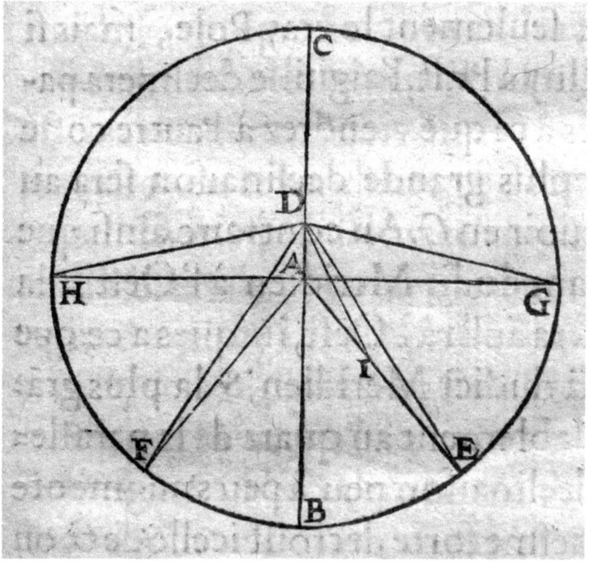

Fig. 8.17 Figure explaining the compass variation in Coignet (1580) (EHC G 16623)

theory was correct, the magnetic variation at Antwerp should be 9°, which corresponds exactly with the observed value.[40] In 1584 or 1585, Mercator sent him a now lost letter concerning this prime meridian.[41]

He goes on to explain the magnitude of the variation according to one's location. Using a polar projection of the Earth (Fig. 8.17), where *A* represents the pole and *D* the magnetic pole, Coignet explains the variation. On the meridian of the Azores *BADC*, the variation is supposedly zero. If one sails East, the compass needle will point east of the pole; sailing west it will point west of the pole. The variation is at its maximum at *G*, resp. *H*, i.e. at 90° and 270° counting from the meridian. If the theory was correct, the maximum should occur slightly beyond G. Given his trigonometrical knowledge Coignet, should have been aware of this.

According to this theory, there are only two places on earth, at a given latitude, where the variation is the same. Hence it is not difficult to calculate one's longitude if given one's latitude, because the two points with the same variation are in different

Cape Verde Islands (where declination was supposedly zero), he arrived at two possible locations for the magnetic pole. Both these magnetic poles are shown on sheet 6 of his 1569 world map. On this model, see for example Jonkers (2008, p. 259).

[40]In Coignet's polar model: $\tan \upsilon = \dfrac{\sin \lambda}{\cos \theta \cot \delta + \sin \theta \cos \lambda}$ in which υ is the variation, λ the geographical longitude, θ the latitude and δ the angular distance between the magnetic and the geographical pole. For Antwerp $\theta = 51°13'$, $\delta = 16°30'$ $\lambda = 26°30'$. The determination of λ is crucial. From the values in the 1575 edition of Apianus's *Cosmographia*, we can deduce that Antwerp should have a longitude of 26°30′ E relative to the prime meridian through the Azores. This yields $\upsilon = 9°1'$. On the polar model, see Johnston (1994, p. 12).

[41]van Durme (1959) after Saminiatus (1599, p. 27).

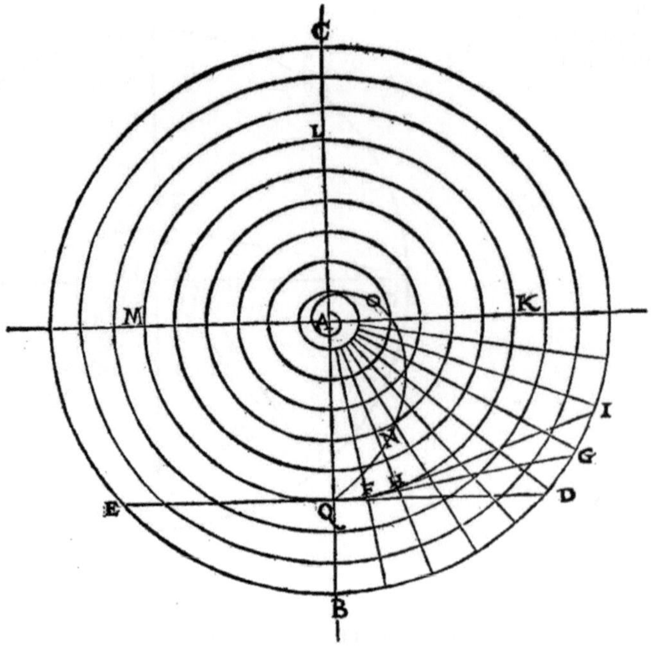

Fig. 8.18 The spiral shape of the loxodrome (Coignet 1580, EHC G 16623)

quadrants. Coignet, however, did not draw this conclusion. Others, such as Plancius (1552–1622) and Stevin (1548–1620), did. They were of the opinion that longitude could be determined by a careful reading of the magnetic variation. However, observations from across the world would reveal that the pattern of variation was highly complex.[42] Among these observers were the Jesuits, Gregory of St Vincent s.j. and his pupils.[43]

In *Instruction nouvelle*, the length of a loxodrome or Rhumb line is calculated and its nature explained. A loxodrome is the line that intersects each meridian under the same angle. A ship keeping a constant compass bearing would sail along a loxodrome. Coignet warns navigators that loxodromes are not great circles (Fig. 8.18). He calculates the length of a loxodrome for a latitude difference of 1°, by approximating the loxodrome locally by a great circle (see Fig. 8.19) at 60° N. However, this length is of little practical value to a sailor, Coignet asserted. The episode exemplifies how mathematicians were trying to resolve the difficulties

[42]Dutch navigators were ordered in 1599 to keep accurate records of the compass variation encountered during their voyages and to pass on this information to the Admiralty, where they could be studied. In this way, the theory of variation could effectively be put to the test. The experiments by, among others, Henry Gellibrand (1635) showed that compass variation is time-dependent and thus unusable. Waters (1978, p. 228) and Davids (1986, p. 79).

[43]Dhombres and Radelet-de Grave (2008, p. 171).

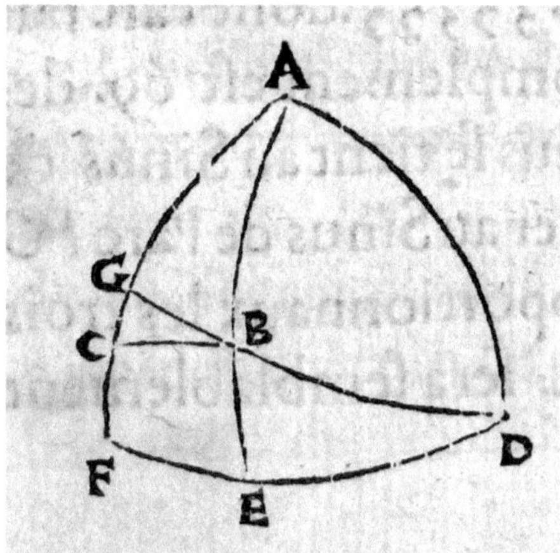

Fig. 8.19 The figure is not to scale (Coignet (1580), EHC G 16623).
B is the North pole, $\angle FD = 90°$, $\angle BE = 60°$, $\angle AB = 30°$
Let AF be the fourth Rhumb (NE) or $\angle EF = 45°$
A ship sails along the Rhumb to C at 61° ($90° - \angle BC = \angle\overline{BC} = 61°$).
Now the question is: what is $\angle AC$?
Now $\dfrac{\sin\angle AE}{\sin\angle AF} = \dfrac{\sin\angle AB}{\sin\angle BG} \Rightarrow \angle BG = 20°42'$ and $\angle\overline{BC} = \angle BD = 69°19'$
and $\dfrac{\sin\angle BD}{\sin\angle BE} = \dfrac{\sin 90°}{\sin\angle FG} \Rightarrow \angle FG = 67°47'$ and $\angle\overline{FG} = 22°13'$
and $\dfrac{\sin\angle\overline{BG}}{\sin\angle\overline{BC}} = \dfrac{\sin 90°}{\sin\angle\overline{CG}} \Rightarrow \angle\overline{CG} = 69°13'$ and $\angle CG = 20°47'$
Now $\angle AC = \angle AG - \angle CG = 1°26'$
If 1° equals 17.5 miles, then $\angle AC = 25.5$ miles

the longitude problem posed. Portuguese scholars had compiled tables with the lengths of loxodromes, in order to be able to draw these curves on terrestrial globes. Calculating the length of a loxodrome is a step on the way to unravelling the secrets of the Mercator chart. Similar calculations of meridian lengths, pushed to their limits, would be performed by Edward Wright, who thereby made Mercator charts practicable.

Medina's and Coignet's book was reprinted three times in Amsterdam by Cornelis Claesz., at a time when the Dutch Republic was rapidly expanding its commercial fleet. The book seems to have had a wide circulation. The remnants of Barentz's copy of *Behouden Huys* were found on Novaya Zemlya in 1871 alongside not only navigational instruments, but also a copy of the Antwerp edition of Medina's and Coignet's book.[44]

[44]Crone (1966) and Heijting (2007, pp. 156–157 and 165).

The book also drew interest from other writers on the subject of navigation. Mercator, for example, owned both the Dutch and the French version.[45] It was mentioned, referred to or cited by such authors as Albert Hayen (1600), Robbert Robbertsz. (le Canu) (1612), Jan vanden Broeck (1610) and Willem Blaeu (1631).[46] The latter copied Coignet's definitions of coastal and oceanic navigation almost literally.[47]

Although the French edition was printed only once, its impact in England was considerable. Thomas Blundiville plagiarized the book and published it as part of his *Exercises* (1594), which would be reprinted eight times. This suggests that, indirectly, Coignet's book greatly influenced English navigators in their quest to rule the waves.

[45]Penneman and van der Gucht (1994, p. 97).

[46]Burger (1908, p. 50) and Waters (1978, p. 161).

[47]Blaeu (1631).

Chapter 9
Mapping the World

9.1 Introduction

The advent of printing gave rise to a specialist field that would become ever more artful as the sixteenth century progressed: cartography. The art of map-making was where mathematics, geography and artistic craftsmanship met. Maps could not be produced without the mathematics to project a sphere onto a plane; the science of geography provided the means to describe landscape features such as rivers, mountain ranges, valleys and plains; and the craftsmanship of the etchers allowed for an undreamt of accuracy as well as artful and fanciful ornamentation. Maps became works of art in their own right (Fig. 9.1).

Nowhere was this art pursued more assiduously than in the Low Countries: the pioneering work of Mercator, Ortelius, De Jode, Blaeu and Waghenaer still captures our imagination today.

9.2 Mapping the Low Countries

As in so many practical applications of mathematics in the Low Countries, Gemma Frisius played a pioneering role in the development of cartography. In 1524, Petrus Apianus published his *Cosmographicus Liber* in the Bavarian town of Landshut. The work met with reasonable success. Five years later, Gemma Frisius's edition of the book was first published by Roeland Bollaert from Antwerp. But it was the revised 1533 edition that became an instant hit, thanks to an ingenious innovation: working paper copies of the instruments described (see Figs. 7.1 and 9.2). Frisius's edition of the *Cosmographia* was reprinted numerous times during the sixteenth century. The book familiarized a wide audience with cartographic and geographic

A. Meskens, *Practical Mathematics in a Commercial Metropolis: Mathematical Life in Late 16th Century Antwerp*, Archimedes 31, DOI 10.1007/978-94-007-5721-9_9,
© Springer Science+Business Media Dordrecht 2013

Fig. 9.1 Detail of a map from de Jode (1593) (MPM A 2880)

principles. At the same time, Frisius's methods were put to practical use by his students,[1] most famously of all by Gerard de Kremer, better known as *Gerard Mercator Rupelmondanus*.

One admittedly rather modest, and local application of the art of cartography lay in the charting of the Seventeen Provinces. The first book on triangulation, the technique that made this possible, was Frisius's *Libellus de locorum* (1533 – Figs. 9.3 and 9.4). Although the work is only 16 pages long, it is a milestone in the history of science. Frisius's triangulation method essentially amounts to the following: given a baseline and the angles from that baseline, or from determined points, one can draw a map of the area. This approach has the advantage that it accurately represents the relative positions of cities, even if the baseline (but not the angles) has been measured erroneously.

Around the time that Frisius published *Libellus*, Jacob van Deventer was making systematic use of the triangulation technique for the mapping of the Seventeen Provinces. In 1536, he produced a printed map of the Duchy of Brabant (which comprised the present-day Belgian provinces of Walloon Brabant, Flemish Brabant, Antwerp and the Brussels Capital Region, as well as most of the Dutch province

[1] See for example Vanden Broecke (2000).

Fig. 9.2 From Frisius and Apianus (1533) (MPM A 2794.2)

of North Brabant), the first such map to be published in the Netherlands. Maps of Holland, Gelre, Frisia and Zeeland would follow. Van Deventer had commercial publishing in mind, but not all of his maps made it into print, as the central government regarded them as militarily sensitive. From 1558 onwards, he would embark on another project, which involved the production of maps of about 260 towns and cities in the Low Countries.[2]

[2]Koeman and Visser (1992), Koeman (1994), Vande Winkel (2008) and Roegiers and vander Herten (eds.) (1994, pp. 26–29).

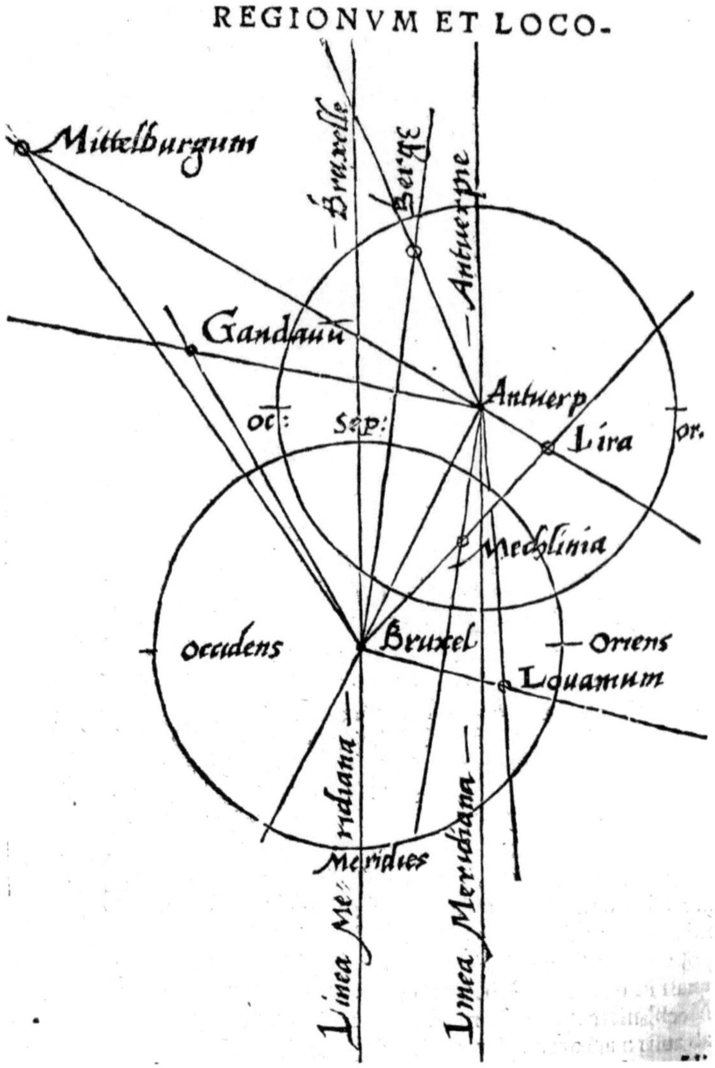

Fig. 9.3 Triangulation in Gemma Frisius's *Libellus de locorum* in Frisius and Apianus (1574a) (EHC G 5048, reprint of 1533)

For the charting of the provinces, the surveyors made use of a number of instruments, including a surveyor's circle. A surveyor's circle, or Holland circle,[3]

[3]A term coined in the nineteenth century, because the Dutch surveyor Dou described it in an early seventeenth century book. The instrument was already described almost a century earlier by Apianus and Frisius.

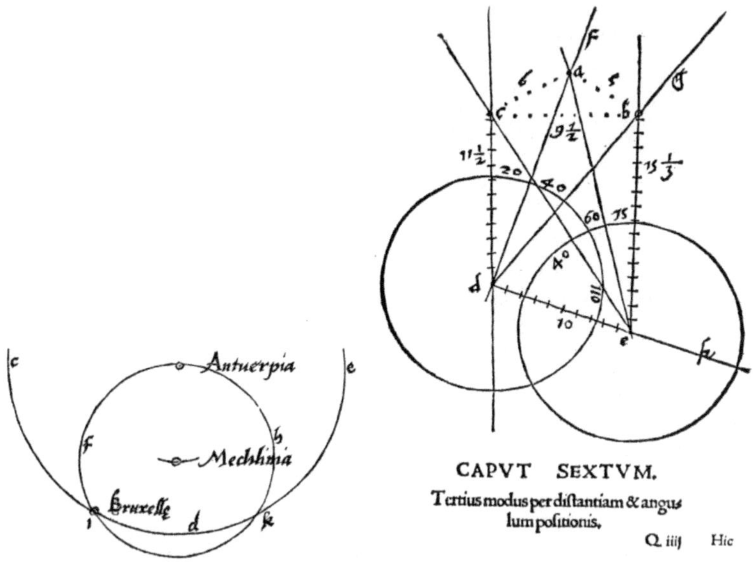

Fig. 9.4 From Frisius and Apianus (1574a) (*left*, EHC G 5048) and Frisius and Apianus (1574b) (*right*, MPM R 24.30)

Fig. 9.5 A Coignet surveyor's circle (*front*). Private collection (See Rasquin (1997a))

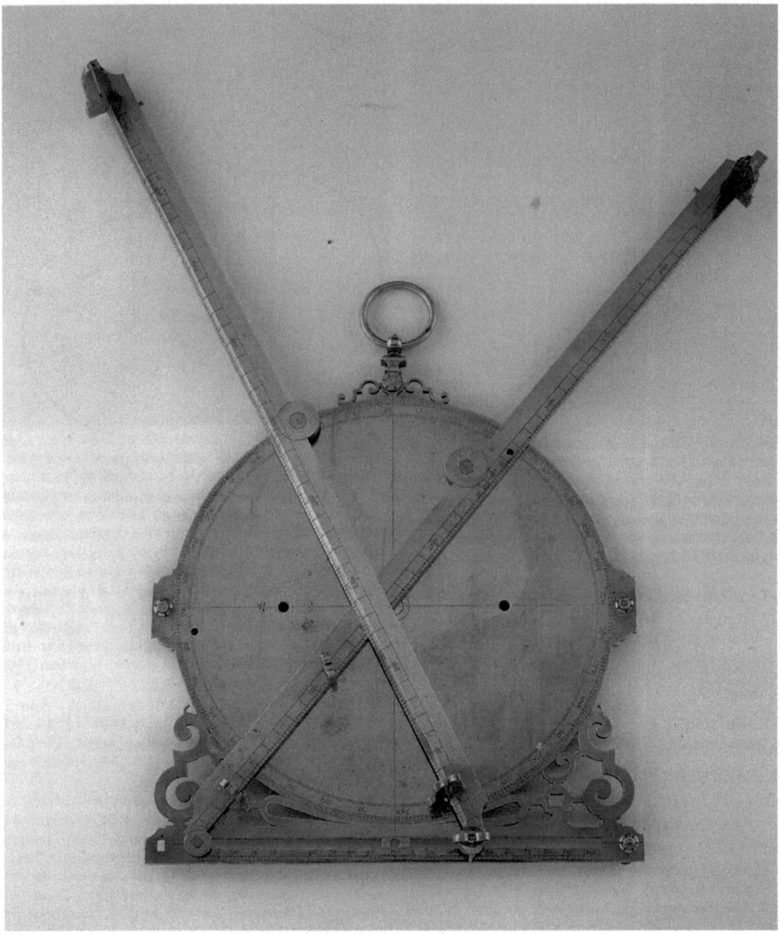

Fig. 9.6 A Coignet surveyor's circle (*back*). Private collection (Rasquin 1997a).
With the alidades with sighting vanes, a direct triangulation is possible. Make a sighting in point *A* with the left alhidade. Move to a point *B* and put the pivot of the second alidade at a point for which the distance *AB* is to the scale of your map. Now make a sighting with the second vane. Transfer the angles to the map

is essentially a round brass disk with a rim that bears an angular scale. It has an alidade for sightings and is used in conjunction with a compass (which is sometimes incorporated into the instrument). Once the instrument is properly aligned, the sighting of a target allows the measurement of the angle between the line of sight and the meridian.

9.3 Mapping the World

Fig. 9.7 Abraham Ortelius (MPM 4-97)

The concept of an atlas took definitive form in the work of Abraham Ortelius.[4] Ortelius was born on 4 April 1527 to the family of a merchant who sympathized with the Lutheran cause. Abraham enjoyed a good education, which included Latin, Greek and mathematics. In 1547, he was admitted to the Guild of St Luke as an *afsetter van caerten* ("seller of maps").[5] Ortelius published his first map of the world in 1564 and, by the 1570s, he had gained great fame in this field. His groundbreaking work, the *Theatrum Orbis Terrarum*, which is considered as the earliest modern atlas, was first published in 1570. The *Theatrum* was an instant success. Distributed by Plantin, it contained maps based on shipowner Gillis Hooftman's collection (see p. 139), accompanied by text in relief print produced by Gillis Coppens van Diest. The *Theatrum* was published in Latin, French and German. After the printing of the text, the copper-engraved illustrations were added by means of an intaglio process. Priced at 6 gl. 10 st., it was an expensive book.

After the Spanish Fury (1576), many of Heyns's pupils had fled home, which meant that he had plenty of time on his hands. When he was approached by Plantin and the engraver Philip Galle to condense the *Theatrum* into the world's first pocket atlas, he jumped at the opportunity. The resulting work would be known as *Spieghel der werelt*.[6] The maps for the first edition were engraved by Philip Galle, who was also the publisher. Again, there were editions in various languages: *Epitome theatri* in Latin, *Spieghel der Werelt* in Dutch, *Le Miroir du Monde* in French, and subsequently also Spanish, Italian and English versions. For the first Dutch edition of 1577, Peter Heyns, the schoolmaster, *wijkmeester* and playwright, wrote the accompanying texts in rhyme. He also edited the French version, which contained abridged texts from the French edition of the *Theatrum*. Hugo Favolius wrote texts in rhyme for the Latin edition.

---

[4]On Ortelius and his atlases, see Imhof (ed.) (1998) and Imhof (1998).

[5]Rombouts and van Lerius (1874, I, p. 159).

[6]It has been suggested by Meeus (2000, 2009) that a letter from Heyns to Ortelius provided the inspiration for this pocket atlas. In it, Heyns describes an imaginary journey through Europe on the basis of the maps in the *Theatrum*. It is this idea of travelling the world and visiting strange lands simply by reading a book that gave them the inspiration for *Spieghel*.

Ortelius's friendship with Heyns would get him into serious trouble in 1587. The Secret Council (a central judicial body) inquired from the City Council about Ortelius's religious convictions. Hendrik de Moy, the city clerk, responded that Ortelius was a Catholic, but that there had been a misunderstanding: a local commander had concluded prematurely from Ortelius's friendship with Peter Heyns, "an adamant Calvinist", that he, too, was a Protestant. Following De Moy's testimony, Ortelius was restored to all his rights.[7]

As previously mentioned, Ortelius kept Coignet informed of mathematical developments in Italy, but otherwise very little is known about their relationship. There is no indication as to when precisely Coignet became involved in mapmaking, but the earliest evidence dates back to the 1590s.

The subjects of map-making and cartography are raised in some of his manuscripts. He wrote introductions to Cornelis de Jode's *Speculum* and Ortelius's 1609 *Theatrum,* and he also edited Ortelius's *Epitome* in the 1600s. Coignet edited the post-1601 editions of *Epitome theatri orbis terrarum*, which were published by Van Keerberghen. He also wrote a new introduction to the book and added 13 maps. Furthermore, he edited the French version, published by Van Keerberghen, and the English edition, published by J. Shawe.[8] The new maps were engraved by the brothers Ferdinand and Ambrosius Arsenius.[9] Among these new maps was one of Japan (Fig. 9.8), accompanied by a literary description derived from Jesuit letterbooks.[10]

Around the same time, Jan Baptist Vrients printed another edition retaining the Heyns-Galle engravings of the 1580s editions of the *Epitome*. Vrients had acquired these engravings, alongside the plates of *Parergon* and *Theatrum*, at the auction of Ortelius's estate.[11] The plates did not mention a scale and they were of an inferior quality than the Arsenius plates. Between 1604 and 1609, however, the Coignet-Arsenius plates were also acquired by Vrients.

Subsequently, in 1612, they were purchased by Jan Moretus at the auction of Vrients's estate. The Moretus inventory mentions both versions, i.e. the Heyns-Galle and Coignet-Arsenius plates, in Latin and in Italian. Moretus printed new frontispieces for these editions and sold them under his own name.[12]

In 1603, Coignet acted as the middleman on behalf of the Archdukes in the purchase of the Spanish version of the *Theatrum* alongside a terrestrial and a celestial globe.[13]

---

[7]Boumans (1952b) and Meskens (1998–1999, p. 103ff.).

[8]Ortelius and Skelton (ed.) (1968, p. vii).

[9]According to Denucé (1941, p. 80), they assisted Frans Hogenberg around 1565 with the engraving of topographical and other maps for *Theatrum*. See also van Ortroy (ca. 1910).

[10]Lach (1977, p. 357).

[11]van de Vijver (1989, p. 255). For a list of plates that came up for sale in this auction, see Arch182.

[12]MPM M324, f55*v*.

[13]Pinchart (1860, I, p. 211) (incomplete citation); Denucé (1912–1913, p. 268); ANL B2800, f695*r-v*. P. van der Krogt suggests that these globes came from the Northern Netherlands, as

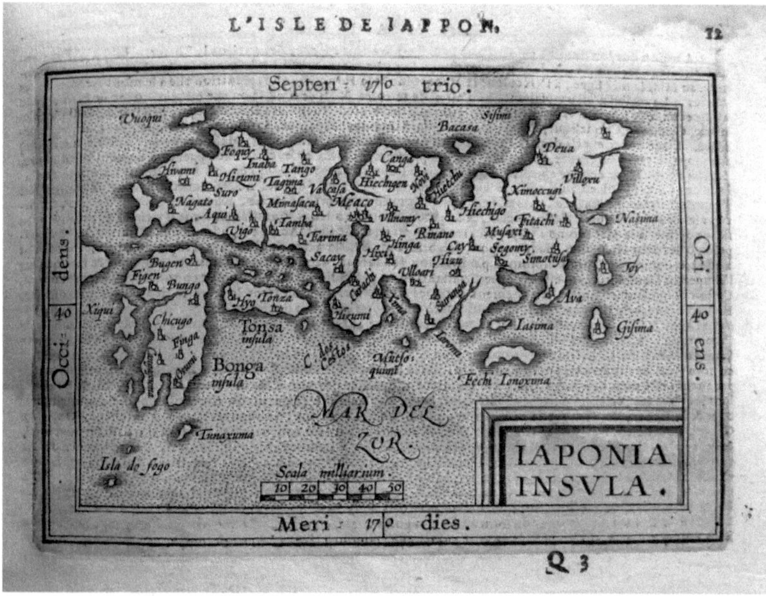

Fig. 9.8 Map of Japan in Ortelius and Coignet (edit) (1602) (MPM 8 190)

In the manuscripts on the sector, Coignet invariably describes the trapezoidal mapping technique. In this approach, a central meridian is drawn, with two perpendicular, equidistant parallels. Two of these parallels, usually the outermost ones, are drawn to true scale. The distance between the outermost meridians – typically at a distance of 30° of longitude – on the map is calculated along both parallels and (drawn to scale) transferred to the map. This gives a trapezoidal outline for the map (see Fig. 9.9).

The circumference of a parallel circle at a given latitude is given by $L = 2\pi R \cos\varphi$, therefore if $\varphi_1 > \varphi_0$ then $L_1 < L_0$. Meridians connecting both parallels are drawn. Coignet uses the trapezoidal mapping technique as an example for the use of the division of sines.

The method lends itself well to the mapping of regions (see Fig. 9.9).

Coignet also wrote an introduction to De Jode's *Speculum Orbis Terrarum* (1593).[14] This introduction contains the first known example of a description of

Vrients obtained permission the following year to buy 20 pair of globes in the "provinces rebelles de Hollande et Zélande" (van der Krogt 1994, p. 165).

[14] van de Vijver (1989, p. 255) and van Ortroy (1914, p. 88).

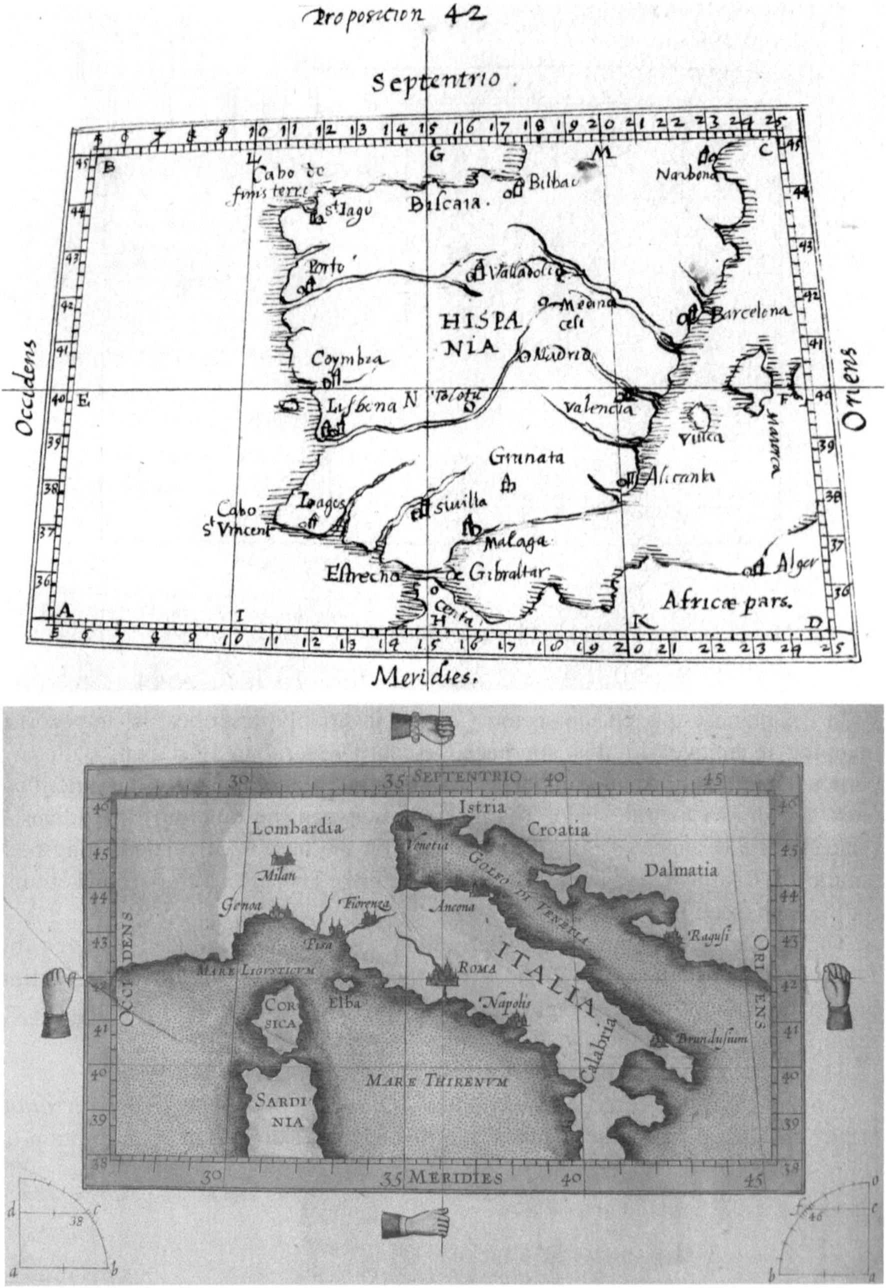

Fig. 9.9 A map of Spain, from Coignet (1618b) (EHC 520935), and a map of Italy, from Ortelius et al. (1612b) (MPMA 2880), created with the trapezoidal mapping technique.
The formulas for the trapezoidal transformation are given by:

$$\begin{cases} x = & R\lambda\left(\dfrac{\varphi\,(\cos\varphi_1 - \cos\varphi_0)}{\varphi_1 - \varphi_2} + 1\right) \\ y = & R\varphi \end{cases}$$

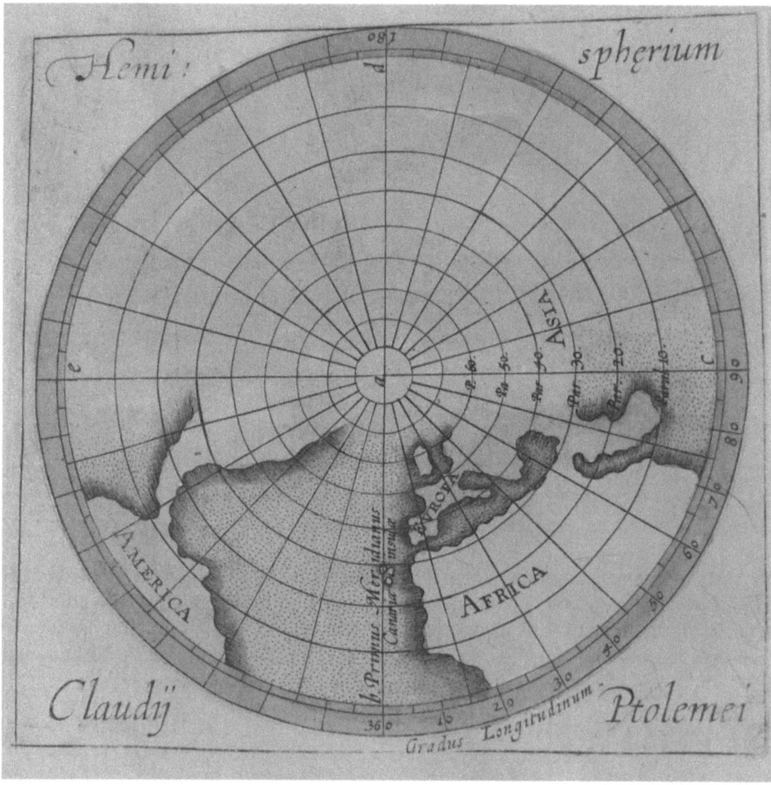

Fig. 9.10 The polar stereographic projection in Ortelius and Coignet (introduction) (1612) (MPM 8-190)

projection methods applied by mapmakers in the production of an atlas. He describes the trapezoidal mapping, the conic mapping and the stereographic projection techniques.

Stereographic projections are of course the projections used in the projection of the heavens on an astrolabe (see Sect. 10.2). Coignet provides details of the polar and the equatorial variants (see Figs. 9.10, 9.12 and 9.11 for an artistic impression).

In a conic projection, parallels are drawn as concentric, equidistant, circular arcs. An additional requirement in this approach is that the distance between two chosen parallels must be to scale[15] (see Fig. 9.13).

[15]On the mathematics of these projections, see Feeman (2002) and Meskens (1998b).

Fig. 9.11 Artistic impression of a stereographic projection (Aguilonius 1613, EHC G 5050). The vignette for Book 6 – as all other vignettes – was engraved by Theodore Galle after designs by Rubens. On Jesuit mathematicians in Antwerp, see Meskens (1997c); on d'Aguilon in particular, see Ziggelaar (1983); on the vignettes, see Meskens (1995, p. 86); on Rubens's designs, see Jaeger (1976)

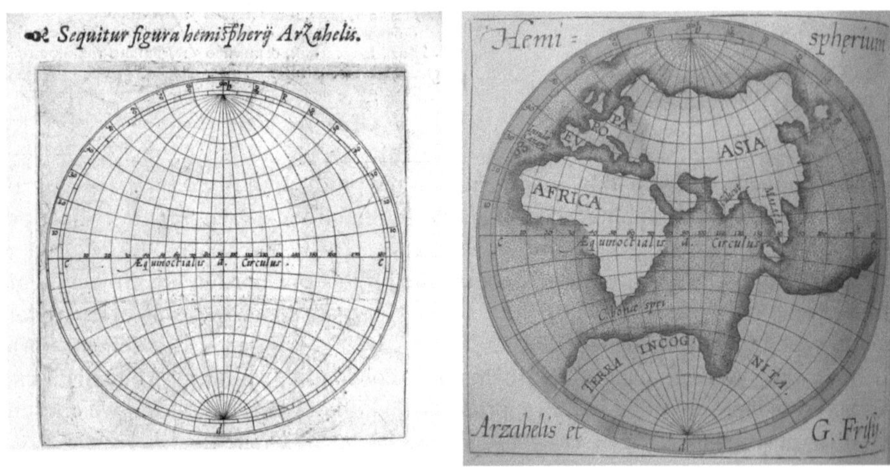

Fig. 9.12 The equatorial stereographic projection, or the projection according to Arzahelis, in de Jode (1593) (*left*) and in Ortelius and Coignet (edit) (1612b) (*right*)

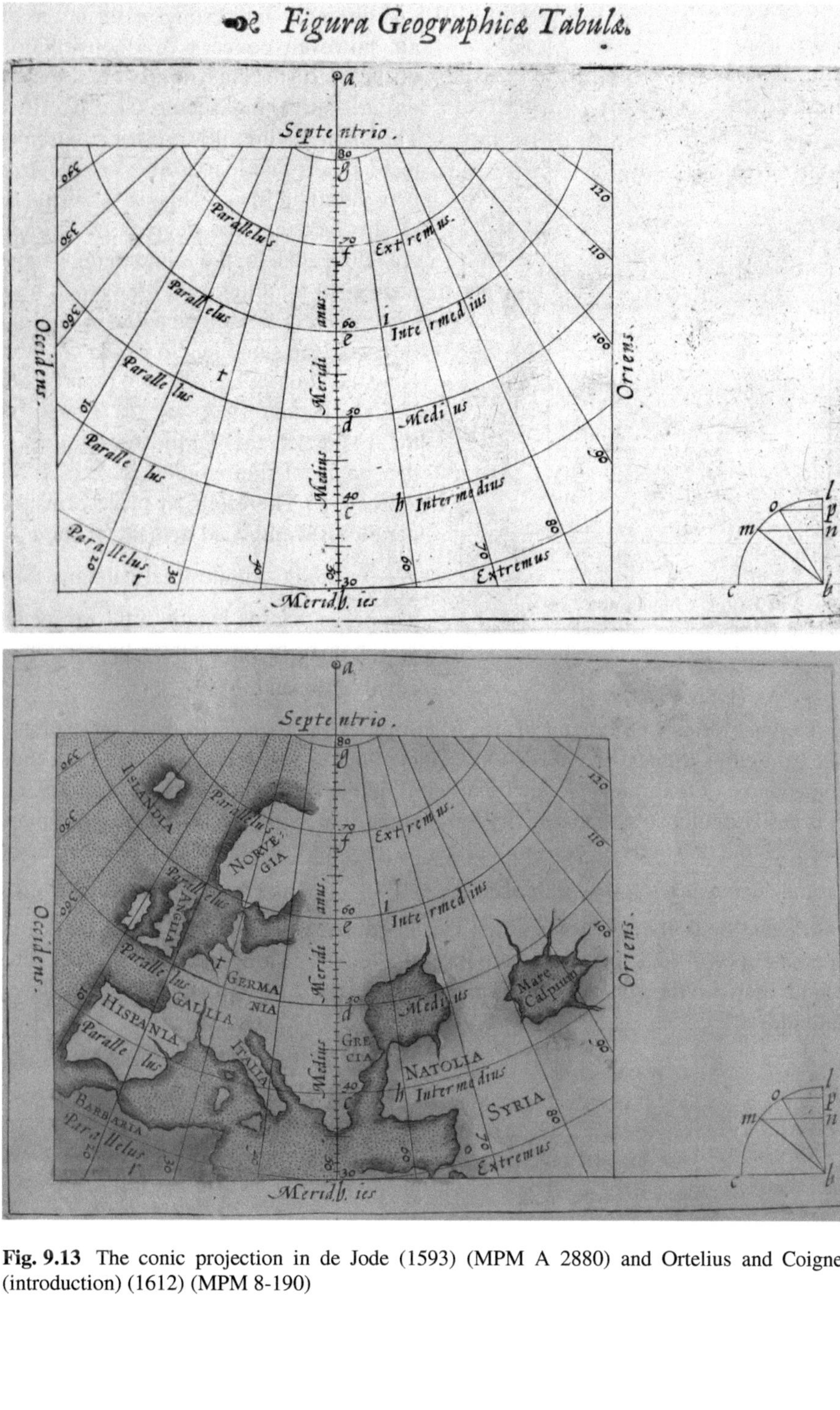

Fig. 9.13 The conic projection in de Jode (1593) (MPM A 2880) and Ortelius and Coignet (introduction) (1612) (MPM 8-190)

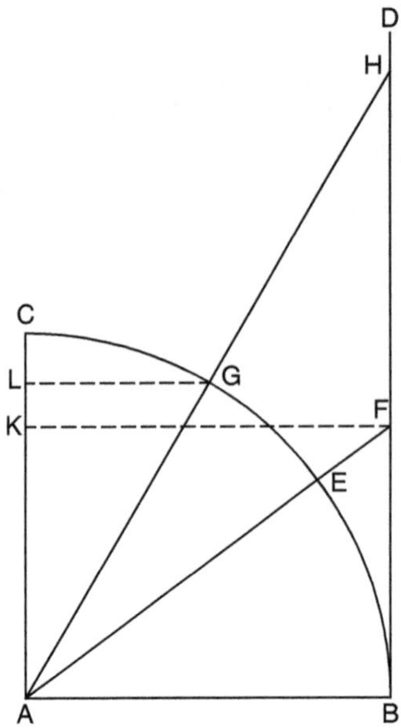

Fig. 9.14 After Coignet (ca. 1604, p. 95) © Paul Tytgat

The most interesting description of all, however, concerns his construction of the Mercator chart in the manuscripts on the surveying sector (see p. 127). The main feature of Mercator's mapping technique is that loxodromes are mapped onto straight lines. This is a corollary of the fact that the degrees of latitude are elongated to the same ratio as the degrees of longitude. On Mercator's map a parallel circle is depicted as a straight line with the same length as the equator. Now the equator has a length (apart from a scale factor) of $L = 2\pi R$, with R the radius of the Earth. At latitude φ the length of the parallel circle is $L = 2\pi R \cos\varphi$. Therefore, in projection, the parallel is lengthened by a factor $\sec\varphi = \frac{1}{\cos\varphi}$. Using infinitesimal calculus, one can prove that the length of the projected meridian from the equator to latitude φ equals $R \ln\tan\left(\frac{\pi}{4} + \frac{\varphi}{2}\right)$.

Coignet refers to Mercator charts as hydrographic maps.[16] To draw such a map, according to Coignet, one needs to calculate the lengths of each degree of latitude for a degree of longitude. Geometrically this amounts to calculating AF in Fig. 9.14. It is easily demonstrated that if $E\hat{A}B = \varphi$ then $AF = AB\sec\varphi$. To draw the map, one uses the divisions of sines on the sector. To determine the length of a degree of latitude between φ_1 and φ_2, calculate $\varphi_m = \frac{\varphi_1 + \varphi_2}{2}$. Take the complement $90° - \varphi_m$ of φ_m and open the sector such that the length between the $(90° - \varphi_m)$-markings (i.e. at $\sin(90° - \varphi_m) = \cos\varphi_m$) equals the length of a degree of longitude. The length between the 90° markings is then equal to the desired length of the degree of latitude.

16 Coignet (ca. 1604), f48*r*.

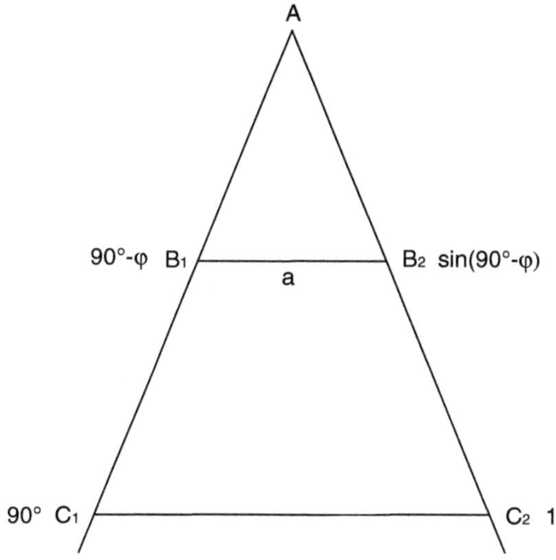

Fig. 9.15 After Coignet (ca. 1604, p.95) © Paul Tytgat

With reference to Fig. 9.15, we see that

$$\frac{B_1B_2}{AB_2} = \frac{C_1C_2}{1}$$
$$\Rightarrow C_1C_2 = \frac{B_1B_2}{AB_2}$$
$$= \frac{a}{\sin(90° - \varphi_m)}$$
$$= a \sec \varphi_m$$

Coignet obviously constructed his map by calculating, for each successive interval, the enlargement factor for the intermediate parallel. Hence, by 1604, he knew how to construct a Mercator chart mathematically. A world map based on Mercator's projection already featured in the 1601 *Epitome* (Fig. 9.16). In the manuscript, Coignet does not mention Mercator, unlike in the *Epitome*, where he acknowledges him as the source for the calculation of "hydrographic tables", which he used in his world map.[17]

For the 1609 edition of the *Theatrum Orbis Terrarum*, Coignet wrote an introduction at the behest of the publisher J.B. Vrients.[18] In 1620, Balthasar Moretus was contemplating a new edition of the *Theatrum.* He asked Ferdinand Arsenius and Arnold-Florent van Langren to produce the engravings, and commissioned Godefroy Wendelin for the texts. The cooperation with Wendelin was unsuccessful and Moretus was also dissatisfied with the work by Van Langren, so that the project was eventually discontinued.[19] But Balthasar had not abandoned all hope, and in 1637 he again considered a reedition of the *Theatrum.* To this end, he asked his brother Theodore Moretus s.j., who was an accomplished mathematician and had studied with Gregory of Saint Vincent s.j., to write a new introduction. In a letter to his brother in Prague, Balthasar reiterates his dissatisfaction with Van Langren's

[17] Ortelius and Coignet (edit) (1602), Introduction (unnumbered page). Coignet refers to Mercator's *Nova et aucta orbis terræ descriptio ad usum navigantium emendate accommodata*, (1569).

[18] Meurer (1991b), Denucé (1912–1913, II, pp. 92 and 250–251) and Bosmans (1928, p. 90).

[19] Denucé (1912–1913, II, pp. 90 and 245) and Bosmans (1928, pp. 89–90).

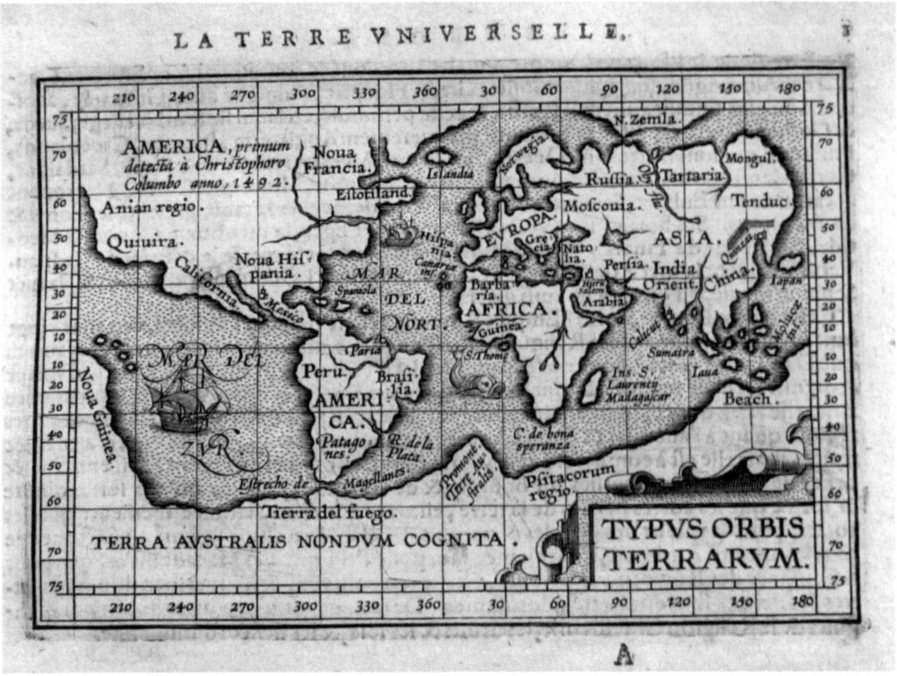

Fig. 9.16 A world map in Mercator's projection in Ortelius and Coignet (edit) (1612b) (MPM 8 190)

engravings while he is very complimentary about Coignet's work.[20] Theodore did not possess a copy of Coignet's treatise. The Jesuit college library in Prague no longer held Ortelius's books either after having been looted by the Germans. Hence, Theodore requested that his brother send him a copy.[21] Balthasar duly obliged. But Theodore's subsequent response to Balthasar's request to write an introduction came as quite a surprise: he wanted to restrict his contribution to a new edition because he felt "Coignet lui-même l'a editée sous une forme assez savante".[22]

The May 1621 edition of *Nieuwe Tijdinghe*, Antwerp's "newspaper", carried an advertisement for a Coignet map featuring a trade route from Antwerp to Milan that was 10–12 days shorter than the regular routes.[23] The ad referred to it as the *Prince*

[20]Bosmans (1928, pp. 89–90).

[21]Bosmans (1928, p. 93).

[22]Bosmans (1928, p. 102).

[23]Meskens (1997b) and Parker (2004, pp. 64–65).

Conduitte, because it lay under the protection of the Archdukes. The map shows a road from Antwerp via Verdun, Besançon and Lausanne, across the Simplon Pass to Domodossola and on to Milan. Hence the entire route passed through Hapsburg territories. However, the war in Germany closed all corridors between Italy and the Low Countries, making the proposed route obsolete very quickly.

Chapter 10
Stargazing

The second half of the sixteenth century saw the transition from a geocentric to a heliocentric worldview. Although the Copernican system was generally received with scepticism, it certainly stirred an interest. With the introduction of the intermediate system of Tycho Brahe, a gradual shift took place towards the Keplerian elliptical model. In a commercial town such as Antwerp, however, merchants turned to astrologers for guidance and predictions.

10.1 Astrology, the Doomed Pseudo-Science

In an era ravaged by war, people sought and found reassurances from astrologers who claimed they could foretell the future. As the stars were believed to influence the body, physicians were more often than not accomplished astrologers. Authors such as Rembert Dodoens took an interest in the stars, not so much for academic reasons, but in order to be able to draw horoscopes and determine, among other things, the right treatment for ailing patients. During the sixteenth and seventeenth centuries, innumerable numbers of almanacs and prognostications were printed. Almanacs and prognostications are essentially different genres, but many intermediate styles existed.

In an almanac, one typically finds a calendar indicating the dates of religious and civil festivals. As the date of Easter, and hence those of other, related festivals such as Pentecost, vary each year, they need to be calculated annually. Also featured were the position of the Sun relative to the Zodiac and the dates of annual fairs or markets. Sometimes they included weather forecasts, much like prognostications did. Some almanacs also provided (pseudo-)scientific information on bloodletting, purging or fluvial tides.

A. Meskens, *Practical Mathematics in a Commercial Metropolis: Mathematical Life in Late 16th Century Antwerp*, Archimedes 31, DOI 10.1007/978-94-007-5721-9_10,
© Springer Science+Business Media Dordrecht 2013

Fig. 10.1 Detail of a clock, made in or around Augsburg, with clear Antwerp influences in the etchings featuring astronomical motifs (KBC-Bank Rockoxhuis)

Prognostications were astrology-based predictions. Since the Council of Trent (1545–1563), the Church had turned on astrology,[1] but despite this opposition a stream of publications continued to flood the market.

Most authors of prognostications and almanacs, which were printed in Antwerp, belonged to the professional circle of physicians. The only mathematician we were able to identify is ... Michiel Coignet, who published almanacs in older age. In the 1616 accounts of the Guild of St Luke, he is mentioned as an *astrologus*.[2] Coignet

[1]This was due to the fact that astrology taught astral determination, which contradicts the Christian doctrine of free will and moral independence. Theologians did not oppose the view that the stars influenced the body, but they did object to the notion that they also influenced the soul. Thomas (1989, pp. 309–310).

[2]Rombouts and van Lerius (1874, p. 11), also MPM Arch 184, f13*r*.

Fig. 10.2 An almanac for 1592, by Jan Verniers, "geometrist" from Ghent (MPM A 2628)

is known to have signed almanacs for 1617 and 1626, so it is quite conceivable that he has also calculated those for the intermediate years.[3] Jan Peetersen from Delft refers to himself as *mathematicus* and *astrologus* (mathematician and astrologer),[4] but unfortunately we possess no further biographical details about him.

[3]City Library of Kortrijk almanacs no. 33 and 39. See also Caullet (1904).
[4]EHC K101982.

All other authors call themselves *Chirurgijn* (chirurgian/surgeon/physician) or *Doctoor inde Medicijnen* (medical doctor).[5] The best known of these authors are Pieter Hasschaert of Armentières, Rembert Dodoens (1517–1585)[6] and Michel Nostradamus (1503–1566).[7] Pieter Hasschaert, who published numerous almanacs with Plantin, also wrote some medical treatises and a book on syphilis.[8] Cornelius Schylander[9] was the author of *Medicina Astrologica* (1577) and astrologically inspired medical treatises.

It has been suggested that late sixteenth and seventeenth-century developments in astrology were a failed (or, more accurately, a doomed) scientific revolution. Although the total number of astrologers continued to grow, the number of astronomer-cum-astrologers actually dropped. Despite the concerted efforts of its followers to found astrology on a solid scientific basis, the division with astronomy was already taking place, it seems.

Astrology as such was beginning to lose its standing and it would not be long before it was relegated to the status of pseudo-science. For printers, it became a source of turnover rather than a field of publication for a learned public. Similarly, mathematicians like Coignet may have seen it as a convenient source of income during hard times.

[5]Adrianus Romanus s.j., who is best remembered as a mathematician, refers to himself as a professor of medicine. Theologians appear to have pointed out to Romanus that his predictions were not in accordance with Church dogma and its position on the free will of individuals. On the astrological work of Romanus, see Bockstaele (1989).

[6]Dodoens probably published almanacs on a regular basis. Attested copies date from 1541 (Kronenberg 1928), 1549 and 1558 (SBA K69133). See also Thijs (1980).

[7]SBA K78526 (3 vols.), almanac and prognostications for 1588, 1592 and 1594 (printed by the Antwerp printer Guillaem van Parijs), not mentioned in Chomarat and Laroche (1989).

[8]Elaut (1960).

[9]Probably born in Alphen-aan-de-Rijn and worked as a doctor in Antwerp in the second half of the sixteenth century. He was the chirurgian of the Beghards. See D. Imhof in Brokken (1994a, p. 168).

10.2 The Astronomical Instruments

The astrolabe (See Figs. 10.4 and 10.6) has been sufficiently described in specialized books.[10] Hence, in what follows, it suffices to provide merely a general outline. The mariner's astrolabe, which is not an astrolabe proper but rather a much simplified device, is described in a separate section (see Sect. 8.4).

The history of the planispheric astrolabe goes back to Classical Antiquity, although there are no surviving items from that period. The earliest examples date from the Golden Islamic Era. These instruments can be subdivided according to the projection used: either a northern or a southern stereographic projection, depending on whether the projection centre is the north pole or the south pole of the celestial sphere (Figs. 10.3 and 10.5). In a stereographic projection, an image is projected onto a plane (usually a tangential plane) from a point on the sphere (usually on the diameter perpendicular to the plane). Such a projection has the property of preserving angles, and of projecting circles onto circles, with the exception of those through the tangential point, which are projected onto straight lines.[11]

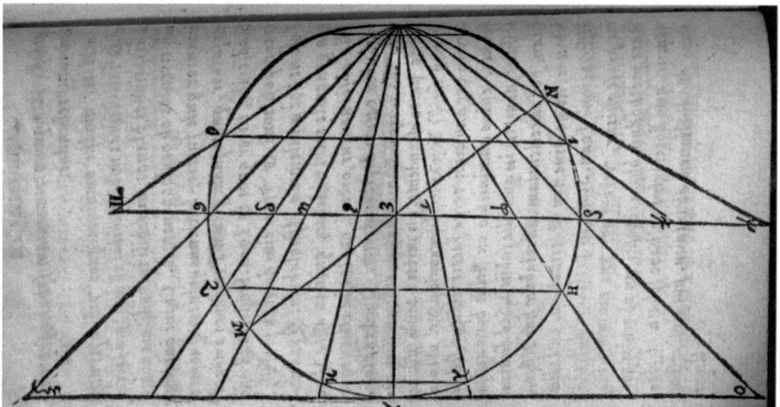

Fig. 10.3 Stereographic projection in Frisius (1556) (MPM A 2429)

[10]See for example Michel (1947), North (1974), Gunther (1976), D'Hollander (1993). A particularly useful book for those unfamiliar with the astrolabe is s.n. (1985).

[11]The map on an astrolabe is not actually a star map. A star map depicts the stars as we see them, whereas in this instance the heavens are projected from outside the celestial sphere onto a plane.

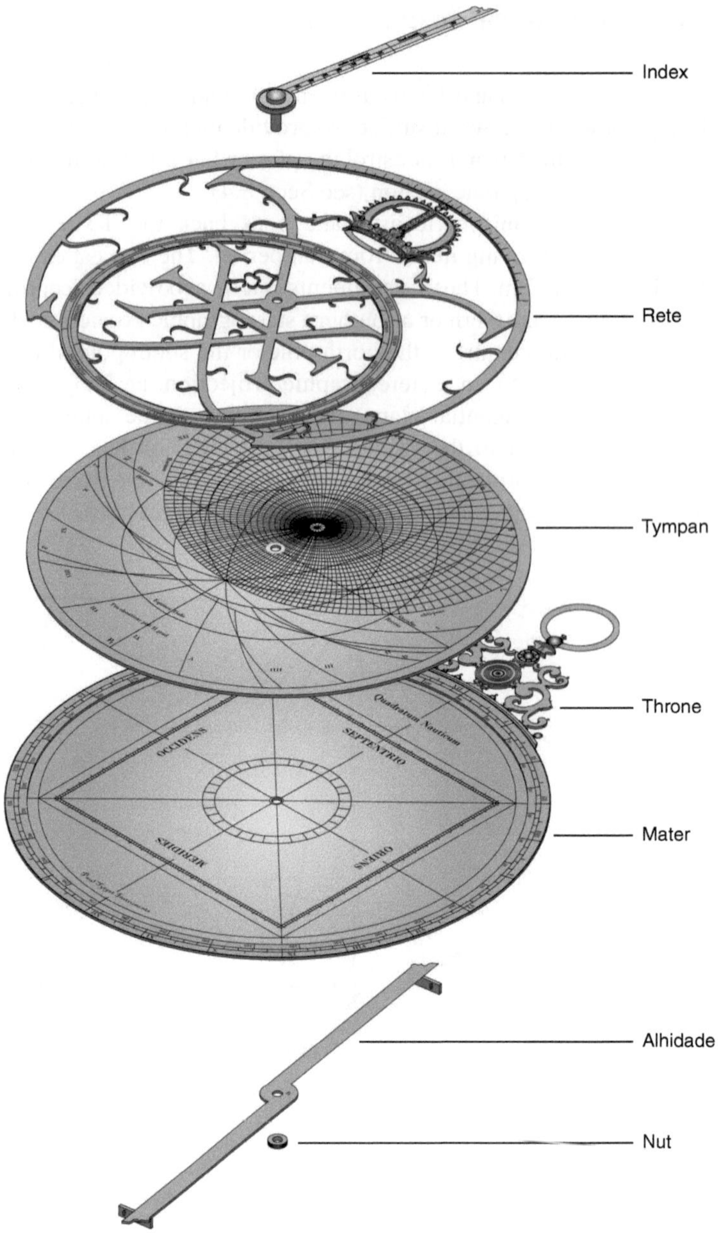

Fig. 10.4 Schematic representation of an astrolabe. The rete is atypical for a Flemish/Dutch instrument and is modelled after the instrument manufactured by the Coignet workshop on behalf of the Archdukes (1618), now in the Museo Arqueológico (Madrid). © Paul Tytgat

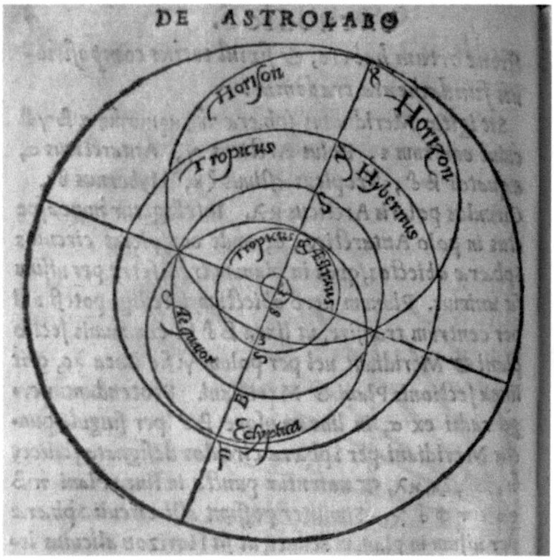

Fig. 10.5 Projection of the principal circles (Frisius (1556), MPM A 2429)

A user will of course be interested only in those celestial bodies he can observe from his position. That is why only those stars are indicated that can, in the course of a year, be seen in the hemisphere with the zenith as a pole.

An astrolabe also bears scales that are directly relevant to astronomical observations, such as a shadow square. Essentially, the device is a sophisticated calculating and observation tool. By manipulating the instrument, one could determine the position of the Sun and the stars. This way, it became possible, for example, to "calculate" the time of sunrise and sundown for any given day.[12]

The armillary sphere represents the Copernican world system (Fig. 10.7). It consists of a series of concentric circular straps. At its centre is a solid sphere, representing the Earth. The spheres of the planets and the outermost sphere of fixed stars are reduced to a number of interlocking rings. Turning one of the rings into the correct position, so as to reflect the actual position of particular body at a given time, would immediately put all the other bodies represented on the sphere in the corresponding positions at that time.

10.3 Copernicanism Versus Tychoism

The Copernican system became known in the Netherlands very early on. In 1548, Rembert Dodoens, the botanist, published a book entitled *Cosmographia*, a digest of the astronomical knowledge of his age, including the Copernican theories. It was republished by Plantin in a slightly altered version in 1584. Dodoens, a student of Frisius's, was well acquainted with Copernicus's views.[13] He did not, however, digress on his system, persisting instead with the age-old Ptolomaic system. Yet the importance of the book should not be underestimated, as this was just a few years after the publication of Copernicus's *De Revolutionibus*.

---

[12]For a detailed description of the use of the astrolabe, see s.n. (1985).

[13]See Dodoens (1548).

Fig. 10.6 Front side and backside of the Coignet astrolabe from the Boerhaave (inv. 3105)

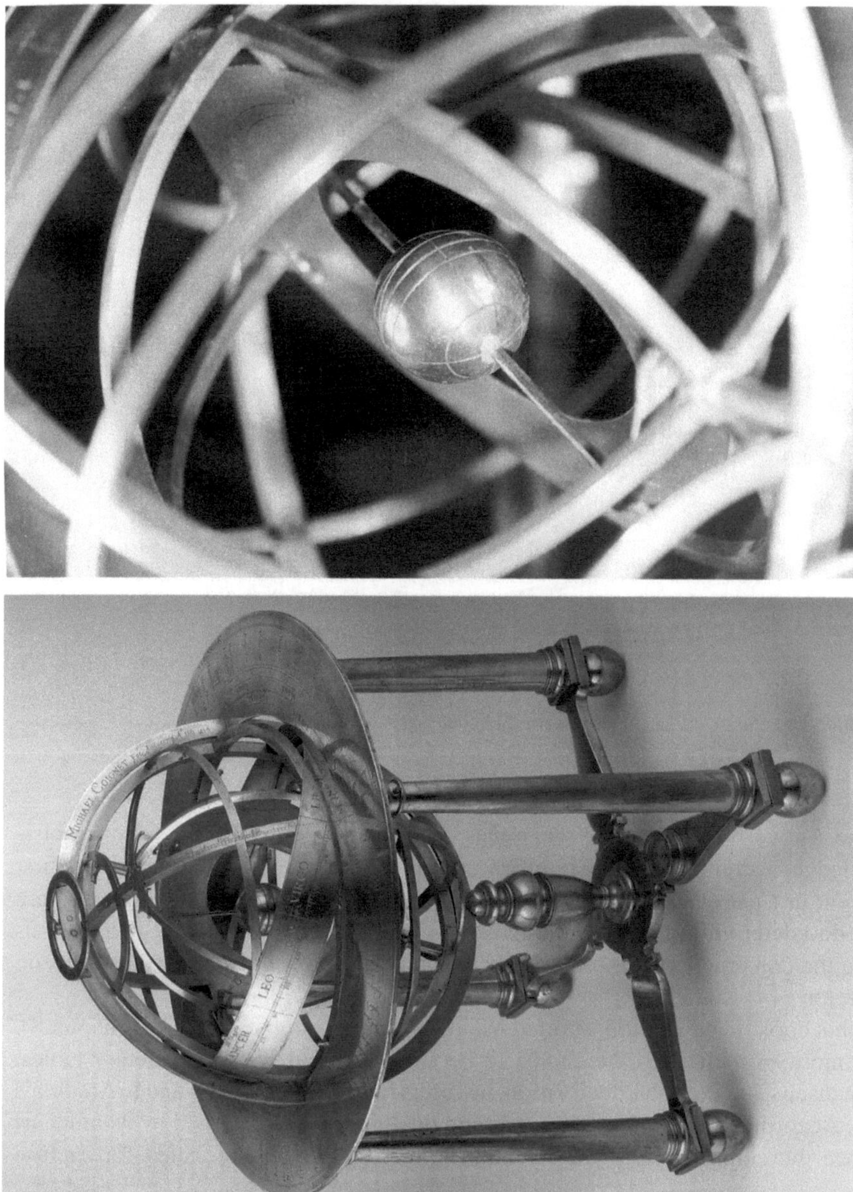

Fig. 10.7 Armillary sphere. Courtesy of Adler Planetarium & Astronomy Museum, Chicago, Illinois

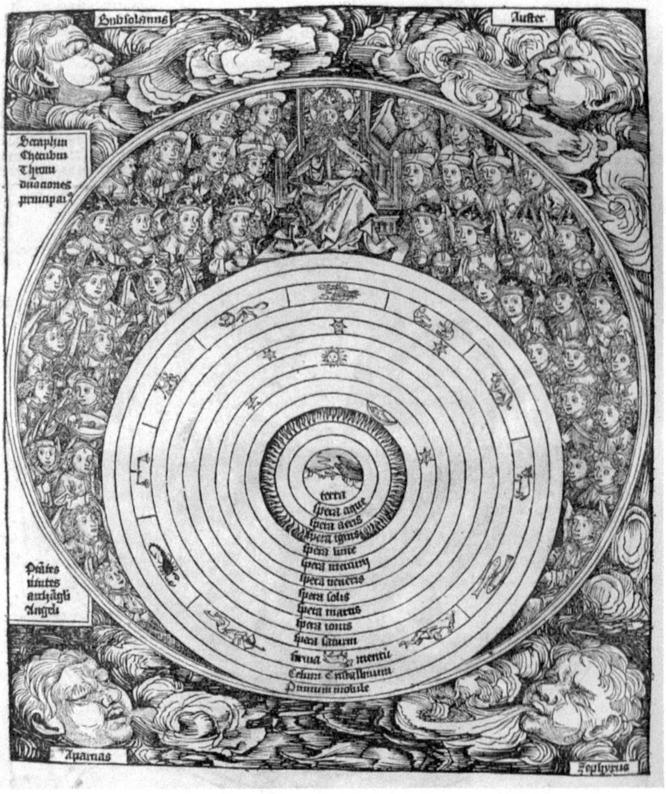

Fig. 10.8 The classical Ptolemaic system (Schedel 1493, MPM R 38.4)

The Antwerp astronomical scene was – unsurprisingly given his trigonometric knowledge – dominated by Michiel Coignet. This interest in astronomy is first apparent in *Cent Questions Ingenieuses* (1573), though the problems proposed are limited to determining one's geographical position given a number of observations and to the construction of sundials. These topics also appear in his manuscripts on the sector.

In his book on navigation, entitled *Nieuwe Onderwijsinghe* (1580, see chapter 8), he is more explicit. Here he claims to be able to calculate new planetary tables. When discussing the solar declination, he refers to a *Graetboecxken* and to Medina's book. According to Coignet, the maximum declination had declined by as much as 5′ since the compilation of the Alphonsine tables.[14] Hence he called for caution

[14]Computations by Swerdlow and Neugebauer (1984, p. 147), indicate that the declination was 23°30.25′ in 1300 and 23°28.20′ in 1525 (no computations are given for subsequent dates). Observational records show that the measured declination was 23°32′ in 1300 and 23°28.24′ in 1525.

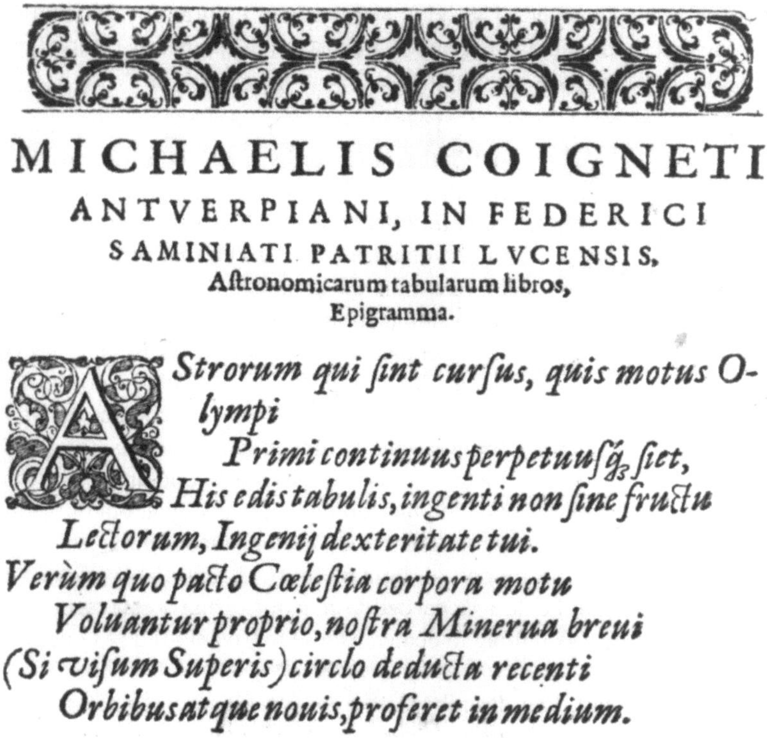

MICHAELIS COIGNETI
ANTVERPIANI, IN FEDERICI
SAMINIATI PATRITII LVCENSIS,
Aſtronomicarum tabularum libros,
Epigramma.

AStrorum qui ſint curſus, quis motus O-
lympi
Primi continuus perpetuuſq; ſiet,
His edis tabulis, ingenti non ſine fructu
Lectorum, Ingenij dexteritate tui.
Verùm quo pacto Cœleſtia corpora motu
Voluantur proprio, noſtra Minerua breui
(Si viſum Superis) circlo deducta recenti
Orbibusatque nouis, proferet in medium.

Fig. 10.9 Epigram (Saminiatus 1599, MPM A 2070)

when using data based on the Alphonsine tables (as in Medina's book). The Alphonsine tables were astronomical tables that had been calculated at the behest of King Alphonso of Castille by a commission under the supervision of Isaac ben Saïd. There is no extant original copy. Peurbach, Regiomontanus and Copernicus compared them with their own observations and noticed discrepancies. On the basis of the results of calculations by Copernicus, Rheinhold compiled tables known as the *Tabulae Prutenicae* (1551). Coignet wrote that he intended to explain the changing values of the solar declination in his *Theoricas Planetarum*, the tables for which he would base on the work of Copernicus and Rheinhold.[15]

In 1599, Coignet's student Federico Saminiati, a descendant from an Italian merchant's family from Lucca, published the misleadingly titled book *Tabulae Astronomica*. In reality, the work deals with sundials and it consists largely in solar declination tables for different latitudes. By means of these tables, sundials

[15]Coignet (1580), f8*r*.

could be correctly installed. The book also bore an epigram by Michiel Coignet mentioning that Saminiati had prepared a treatise on the planetary orbits (Fig. 10.9). This treatise, however, was never published.

Fig. 10.10 Ottavio Pisani (Pisani 1613)

In 1613, Ottavio Pisani (Fig. 10.10) published the only other noteworthy book on astronomy to appear in Antwerp. Entitled *Astrologia seu motus et loca siderum* (1613), it is a lavishly illustrated book, with huge volvelles (Fig. 10.11), that provided the means to determine the positions of the planets. By the time of its publication, however, it was out of date. Pisani, like Coignet, corresponded with Kepler and Galileo, but the latter was very disdainful towards him.

He made it known to Galileo that he had made observations of Jupiter and its moons, suggesting that he had at his disposal very good quality instruments. Galileo himself had lamented that, out of the hundred or so telescopes he had made, only ten were good enough for observing the moons of Jupiter. In 1613 and 1614, Pisani wrote to Galileo to tell him that he had succeeded in manufacturing a pair of binoculars.[16] As with his book, he was willing to dedicate these to Cosimo II de Medici, in return for payment.

Kepler, for his part, had described different types and combinations of lenses and had published the results of his experiments in *Dioptrice*. Hence it comes as no surprise that Pisani wrote to him to seek his advice on the manufacture of binoculars adapted to both eyes.[17] Kepler responded that he had attempted this, but that all his efforts had failed, which seems to have depressed Pisani.[18] Nevertheless, he retained his interest in optics and once again sought Kepler's advice, this time on the curvatures of the lenses in a telescope.[19] Coignet had been involved in publishing atlases from the 1590s onwards. Although he was familiar with both Copernicus's and Tycho Brahe's worldviews, the description he gave in the introduction to these atlases remains Ptolemaic: the Earth, a sphere consisting of water and land embraced by air, is the centre of the universe. The upper regions consist of nine spheres. The ninth sphere or prime mover rotates at constant angular velocity, which is imparted on the other eight spheres. The lowest sphere is that of the Moon, which is the highest anything coming from Earth can reach.

An indication of Coignet's conception of the universe comes from an unusual source, namely a dedication he wrote in Gillis Anselmo's (1575-?) *Liber Amicorum*

[16]van de Vijver (1989, p. 254) and Galileï (1968, I, pp. 564–565), dd. 5 September 1613.

[17]Kepler (1938–, 17, pp. 76–77), dd. 17 October 1613.

[18] Kepler (1938–, 17, p. 98).

[19] Kepler (1938–, 17, pp. 244–246).

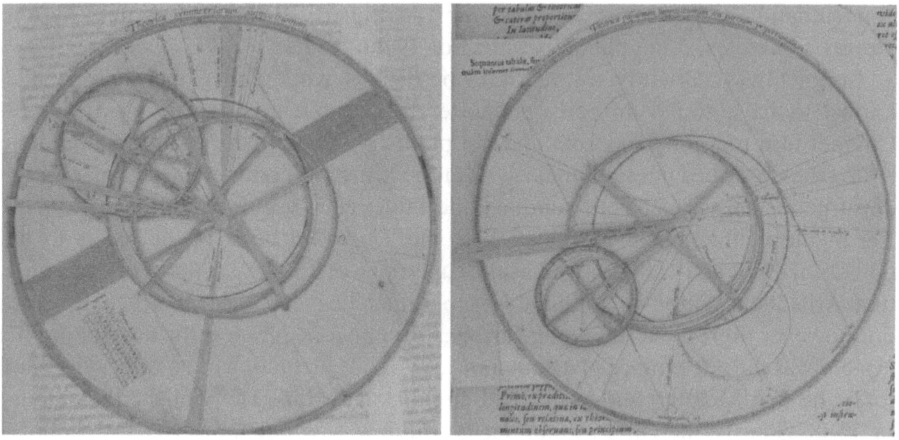

Fig. 10.11 Two pages from Pisani (1613) (EHC G 30235) with moveable parts indicating the positions of the planets

of 21 October 1601.[20] He is the son of Antonio Anselmo (1536 – after 1608), a merchant in woollen who was married to Johanna Hooftman, the daughter of shipowner Gillis Hooftman (see p. 139). In 1580, he was a representative at the States-General that proclaimed the *Act of Abjuration*, a de facto proclamation of independence. He left Antwerp in 1585 and settled in Bremen.[21] The Coignet astrolabe at Museum Boerhaave carries the crest of the Anselmo family on the throne (Fig. 10.6).[22]

In the aforementioned *liber*, Coignet drew a figure representing the Tychonic system, to which he added the comment:

> *Systema Vranicum, iuxta quod noue Caelestum Orbiū hypotheses, M. Coignéti phęnomenis congruentes*[23] *excogitate sunt*
>
> This is the system of the heavens, leading to the new hypotheses of the heavenly orbs, proposed by Michiel Coignet, which ties in with the observed phenomena

Although Coignet claims this geo-heliocentric system to be his own, it was first proposed by Tycho Brahe.[24] In his model, the sun and moon orbit an immobile Earth, while the five planets orbit the sun. As a system, this was acceptable to those who deemed the Ptolemaic system to be untenable, but also questioned the Copernican system on theological grounds. Tycho Brahe's model would gain ground in the early seventeenth century after the invention of the telescope and the Catholic

[20] Anselmo (1594–1602), f30*v*–31*r*.

[21] ARA Raad van State en Audiëntie, 2638, los stuk, dd. 27 September 1608.

[22] See Meskens (2001).

[23] congruentibus, ibus struck through.

[24] On Tycho's system, see Schofield (1961); on Tycho, see Thoren (1990).

condemnation of the Copernican system.[25] As Fig. 10.12 shows, Coignet's system closely resembles, if not plagiarises, that of Tycho Brahe.

Coignet was certainly aware of Copernicus's work and Copernicus is also known to have found an early supporter in Gemma Frisius.[26] Why, then, did Coignet turn to Tycho's system and not publish his own theory? In the atlases he edited, he describes the Ptolemaic system, possibly for commercial reasons. In the construction of astrolabes or sundials, the Ptolemaic system would have sufficed. Yet in the mathematical introduction to the *Theatrum* of 1612, he refers to the comets of 1572 and 1600, as well as the supernova of 1604, to support the argument that Aristotle's notion of immutable superlunar heavens was incorrect. He does not however digress on this topic further.[27] Tycho's inspiration for developing his system was the appearance of a comet in 1577. The calculation of its orbit suggested to Tycho that Mercury and Venus had circumsolar orbits. He published his theory in 1588. Shortly thereafter, a similar worldview was put into print by Nicholas Reimarus Ursus.[28] We know Coignet was a capable, though probably not a prolific, observer of the sky. Even before Tycho's publication, he occupied himself with drawing up astronomical tables, most likely using Copernicus's system. So why did he turn to Tycho's system? It is unlikely that the arguments of the scholastic Aristotelians would have convinced a man so firmly rooted in practice. The more plausible explanation is that, in addition to theological considerations, Coignet became unconvinced of the tenability of Copernicus's system, not least because of the absence of a parallax. Furthermore, it would have been expedient to adhere to Tycho's system in courtly surroundings where the ideas of the Counter-Reformation prevailed. For that matter, Coignet was not the only one to adopt this position: in 1618, Thomas Fyens and Libert Froidmont vigorously condemned the Copernican system. Froidmont would subsequently become a staunch advocate of Tycho's system.[29]

For about 20 years, Coignet maintained that the publication of his astronomical work *Theoricas Planetarum*, containing tables, was imminent, but it is unclear whether it was ever published or even whether manuscript copies of it existed. In *Idea Mathematica* (1593), Adrianus Romanus s.j. writes that Coignet was working on a new theory of the "second movers".[30] The term "second movers" refers to the spheres on which the moon, the sun and the planets and sometimes the fixed stars revolve. In a letter to Christopher Clavius of 3 October 1595, Romanus asserted that Coignet had not finished his *Theorica* yet.[31]

---

[25]Mosley (2007, p. 28).

[26]Waterbolk (1974) and Gingerich and Westman (1988, pp. 42–43).

[27]Ortelius and Coignet (introduction) (1612), fE*r*.

[28]Gingerich and Westman (1988, p. 50).

[29]Schofield (1961, p. 169).

[30]van Roomen (Romanus) (1593, f**jv). "Unum hoc addere libet, cum in doctrina secundorum mobilium restaurande non segitur versari, ita ut brevi novas nobis theoricas polliceatur".

[31]Bockstaele (1976, p. 119).

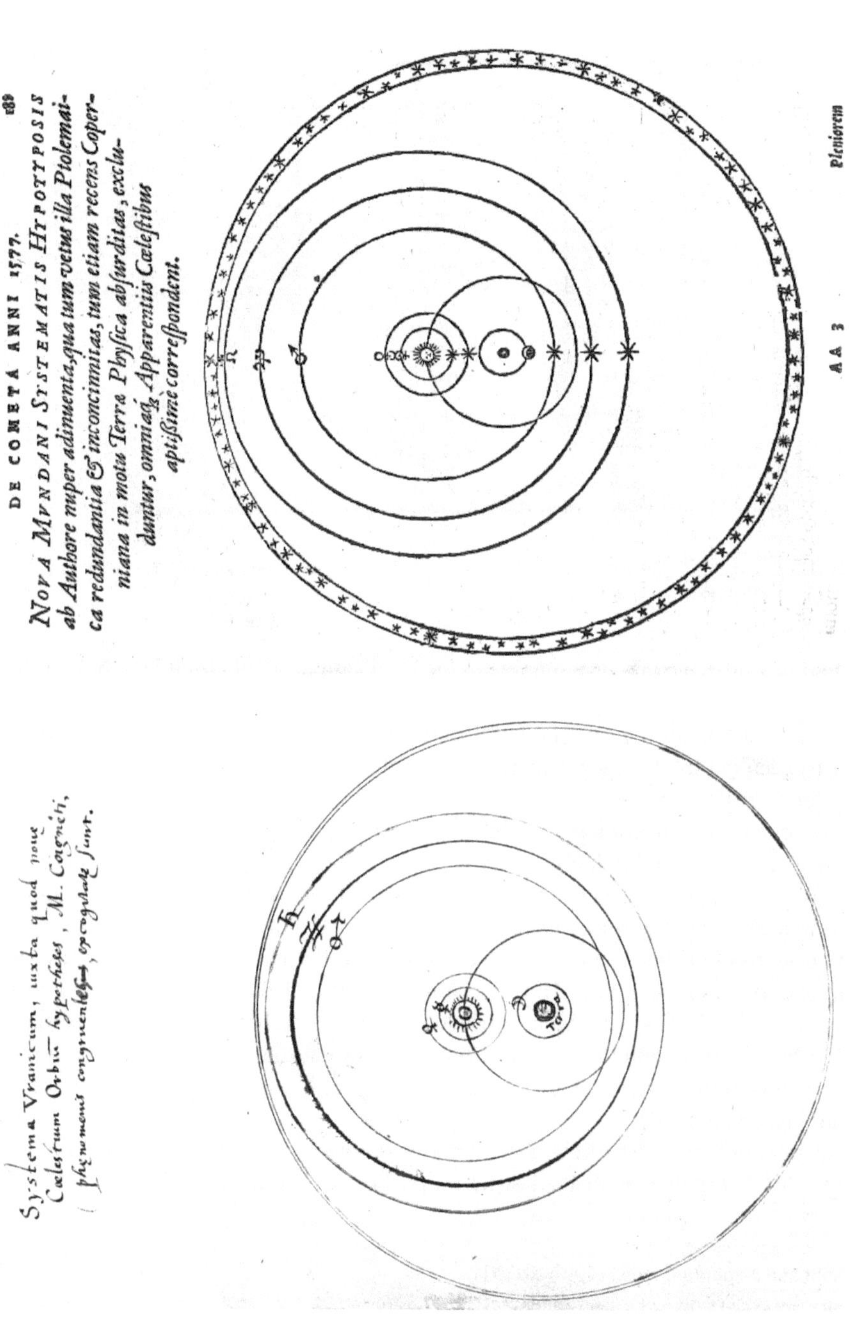

Fig. 10.12 Coignet's system (*left*, Koninklijke Bibliotheek Den Haag BDH 71 J 57) and Tycho's system (*right*, EHC A 125)

Fig. 10.13 Prognostication by Joannes Portantius following the appearance of a comet in 1577 (Portantius 1577, MPM R 63.8 I 75)

Clavius must have taken an interest in the book and probably inquired about it. In February 1603, John Hay s.j. informed him that Coignet was still working on it, but that he was confident it would be published before the *Frankfurter Herbstmesse*.[32] His astronomical treatise was apparently completed by 1606, because he mentions it in a letter to Kepler.[33]

The appearance of comets and other unexpected celestial phenomena usually caused a stream of publications (Fig. 10.13). And certainly this was the case in 1605, when a supernova appeared in the skies. The content of these new books was however very different. In addition to the customary astrological texts that claimed to foretell the future, there were now also books that debated the nature of the phenomenon observed. The supernova became visible in October 1604, just before sundown. It was not visible subsequently in December until after Christmas. Galileo hoped to be able to provide evidence of a parallax shift, thereby proving that the Aristotelian dogma that the heavens above the sphere of the moon were immutable was not true.[34]

The phenomenon also drew the attention of other scientists, notably Michiel Coignet, who, together with Paul Arnerio, purportedly wrote a book on it entitled

[32]Baldini and Napolitani (eds.) (1992, no. 204).

[33]Kepler (1938–, 15, p. 544).

[34]Drake (1976, p. 117ff.).

Discorso sopra la nuova Stella (Padua,1605).[35] We know of no copy of the book, but according to Alberi the authors assert that the phenomenon took place outside the sphere of the Moon. This would be consistent with Coignet adhering to the Tychonic worldview. In this system, the comet could be placed within the void between the Moon and Venus.

In 1618, the appearance of three comets again triggered a stream of treatises. As the last of these comets was the most impressive, it is often referred to as *the* comet of 1618. It was first seen in mid-November, emerging from the solar glare as a tail projected above the horizon. By mid-December, the tail had assumed impressive proportions. It was last seen on 22 January 1619.[36] Gregory of Saint Vincent s.j. and his students made observations and reported to Rome.[37] The Antwerp Jesuits observed the course of the comet using telescopes.[38] Gregory had his pupils defend theses on the nature of the comet in a series of highly successful lectures that drew large audiences.[39] While the appearance of a comet might be a once-in-a-lifetime experience, the Father General warned Gregory that his pupils should adhere strictly to the Jesuit rules. As they were at an impressionable age, no exception should be made.[40] Meanwhile in Rome, Horatio Grassi s.j. anonymously published *De tribus cometis annus MDXVIII*. He argued for the celestial nature of comets by citing parallax measurements made in Rome and Antwerp "scarcely ever exceeding one degree".[41] Other foreign expert observers also seem to have favoured the Antwerp observations.[42] Already during the Christmas holiday of 1618, the Jesuits at the Collegio Romano organized discussions on this subject. Grassi proved, by comparing observations at different locations, that the comet was far more distant than the moon, thereby shattering the Aristotelian doctrine that comets were atmospheric phenomena. The Jesuits, staunch Aristotelians though they were, turned out not to be his fiercest opponents; Galileo was.[43]

---

[35]Alberi (1853, p. 398). Alberi states that the format of the book was quarto, so he must have seen it with his own eyes. Despite our best efforts, we have been unable to locate a copy. Carli and Favaro (1896) mentions *Discorso sopra la stella nuova comporsa l'ottobre prossimo passato, dell' eccelentissimo astrologo et medico Astolfo Arnerio Marchiano*, Lorenzo Pasquati, Padua, 1605. Might confusion have arisen with regard to this book? *Astolfo Arnerio* is thought to be a pseudonym for Spinelli Girolamo, a student of Galileo's.

[36]Yeomans (1991, p. 51), also Jorink (1994). According to A. Tacquet, the tail measured 90° at its maximum (Tacquet (1707, p. 231)).

[37]Puteanus (1619, p. 168); ABML Fonds C. Droeshout, t.III, p. 246.

[38]van Nouhuys (1998a, p. 240) and Riccioli (1651, pp. 101, 102, 111).

[39]See Bosmans (1903), van Nouhuys (1998a, pp. 240, 252) and (van Nouhuys, 1998b, p. 28).

[40] ABML Fonds C. Droeshout, t.III, p. 247.

[41]Yeomans (1991, p. 57).

[42]van Nouhuys (1998a, p. 240).

[43]Ziggelaar (1995, p. 106); on the heated debate between Galileo and the Roman Jesuits, see also (Yeomans, 1991, pp. 51–52). Riccioli (1651), after careful comparison of observations of the 1618 comet, concluded that it remained "unproven" that comets were superlunary, though it was deemed highly likely that they were.

The fact that Coignet himself observed the skies is apparent from his letter to Kepler in March 1606.[44] In it, he communicates his observations of the solar eclipse of 12 October 1605. Kepler had previously received observations made in Antwerp by Gillis Martini,[45] who refers to himself as *Advocatus et Mathematicus*, which leads us to suspect he engaged in some legal astrology. Nothing else is known about Gillis Martini. Perhaps he was a member of the Italian merchant family based in Lucca and with branches in Antwerp, Vienna and Bilbao.[46] Observations of solar eclipses were deemed important, because from the time differences one could infer longitudinal differences. Kepler had also received observations made in London and in Emden, but his calculations yielded implausible results. So he called on Coignet for assistance. In Kepler's opinion, the Antwerp and London observations were in agreement, but if they were correct, the longitudinal difference between Prague and Antwerp would have to be a mere 5°45′, while in fact it is about 10°.

Although we know of no texts on astronomy by Coignet, he must have enjoyed a good reputation in the field. Vendelinus (1580–1667) relied on him to correct his astronomical tables. Coignet asked him to provide values for the years 1000, 1500 and 1600, which he would subsequently check.[47] Godefroy Wendelin, or Vendelinus, a school principal in Herk-de-Stad, was ordained as a priest in 1620. He collected observations of eclipses on the moons of Jupiter in order to test Kepler's second law. After Galileo's trial, he became a fierce opponent of the heliocentric theory. Vendelinus was also the discoverer of the relation between the period of a pendulum and temperature. During a sojourn in Provence, Vendelinus became acquainted with Nicolas-Claude Fabri de Peiresc. The latter is known to have journeyed to Antwerp in 1606, but his contacts in the city remain unclear.[48]

---

[44] Kepler (1938–, 15, p. 543).

[45] Kepler (1938–, 15, pp. 287–288), dd. 1 January 1606.

[46] Baetens (1976, 1, pp. 197 and 222).

[47] Bosmans (1909b, pp. 76–77).

[48] See Lebègue (1943).

Chapter 11
Ballistics and Fortifications

11.1 Introduction

From 1540 onwards, the age-old Hapsburg-Valois conflict, which up to then had been limited to Italy, spread to the Netherlands. François I began to build Italian-style fortresses along the Franco-Netherlandic frontier. Charles V responded by ordering the construction of his own series of forts against the French. This highlighted the fact that, while the Low Countries, with their fortified towns, canals, waterways and dikes were virtually impenetrable[1] to the French, France was relatively exposed to attacks from the Netherlands. Had it wanted to, Spain could have invaded France from the north almost unimpeded.[2]

At Saint-Quentin, Philip forced France's hand, resulting in the peace treaty of Cateau-Cambrésis (1559). By now, however, civil unrest had spread throughout the Low Countries. The uprising was countered by the incorporation of new citadels into the city walls, in order to keep the towns in check.

Sixteenth-century architecture was a field of knowledge that was very much open to mathematization.[3] Nowhere is this more apparent than in the theoretical approach to building fortresses. Indeed, in architectural manuals, we find ample evidence of geometric constructions that are entirely absent from arithmetic books.

Artillery had become much more potent than it had been in previous eras, rendering mediæval city walls redundant: they would simply be pulverized by the firepower of modern cannons. And once its defensive walls were breached, a city could of course be stormed. In Italy, with its many rivaling principalities, a new type of defence was developed: the stronghold. Soon, the Italians became so experienced at the construction of such defences that their engineers were in great demand throughout war-torn Europe.

---

[1] A fact which, half a century later, the Spanish would also be faced with.

[2] See Roosens (2005) and Martens (2009)

[3] See Bennett and Johnston (1996).

A. Meskens, *Practical Mathematics in a Commercial Metropolis: Mathematical Life in Late 16th Century Antwerp*, Archimedes 31, DOI 10.1007/978-94-007-5721-9_11,
© Springer Science+Business Media Dordrecht 2013

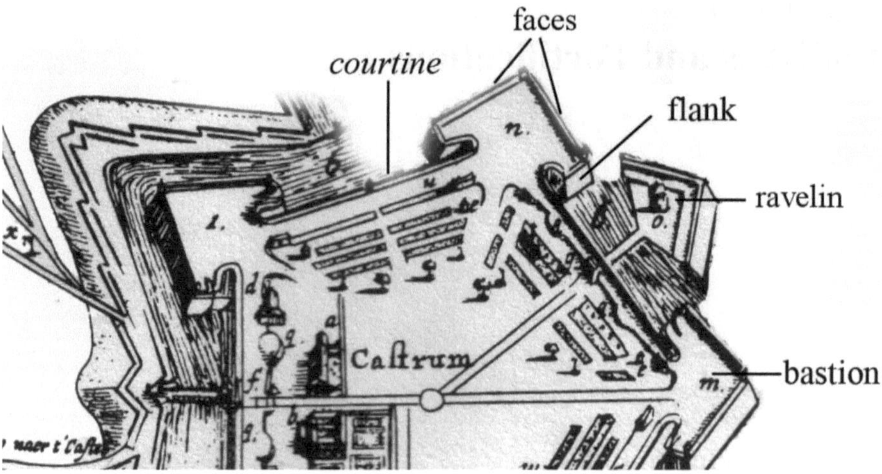

Fig. 11.1 Terminology (adapted from SAA BIB 1932)

The stronghold or bulwark was a line of defence with walls that, ideally, were built along a regular polygon, with bastions at the vertices (See Fig. 11.1 for terminology). It was surrounded entirely by a moat. The bastions consisted of two sideways-facing "flanks" and two outward-facing "faces". These bastions served a double purpose. The meeting of the two faces of the bastion in an angular salient eliminated any dead ground, while the provision of a wide flank enabled the defender to bring a heavy cross-fire to bear along the ditch. Moreover, the long straight parapets enabled the defender to mount a large number of cannons. The design of the bulwark was such that every part of the structure could lend support to another part in case of an attack.[4]

Constructing such bulwarks was expensive, though, so that the Dutch insurgents developed another type of defensive line, in which the actual defences were preceded by an earthen wall and a ditch. On top of the earthen walls, *half moons* were constructed to provide additional support to the bastions. Such walls offered the further advantage of being virtually impenetrable to artillery fire.

Most observers were of the opinion that it was all but impossible to breach a bastionized bulwark and that it could only be forced into submission by a prolonged siege. And this was of course precisely the tactic that the Spanish employed in their campaign in the Low Countries. However, the geography of the Low Countries gave fortresses another advantage. The relatively high water table made it possible for large tracts of lands to be flooded. This in turn offered two benefits: the besiegers were kept at bay and the city could be provisioned by vessels (Fig. 11.2). Hesitation

[4]On the bulwark and its defence, see Duffy (1996, p. 25); on its application in the Low Countries van den Heuvel (1991), Westra (1992).

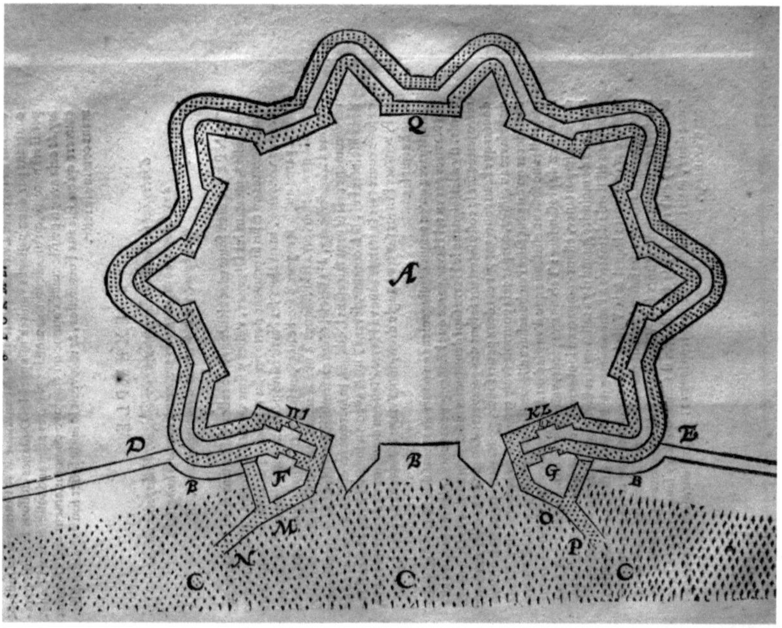

Fig. 11.2 An example of a fortress that can be reached by water (Stevin (1617), EHC G 10414:d)

on the part of Philips of Marnix, Lord of St Aldegonde during the siege of Antwerp (1585) meant that the order to breach the dikes and flood the surrounding land came too late, resulting in a quick surrender of the city.

11.2 Ballistics

The bulwark had been developed in response to more powerful artillery. With this increase in firepower came the desire for greater accuracy, in order to maximize the impact of the ball. This was the century that the science of ballistics was born. The first, hesitant, steps had been taken by Leonardo da Vinci, but more elaborate theories were developed by scholars such as Tartaglia, Galileo, Newton and Huygens. Coignet is a small footnote in this story.

In 1537, Tartaglia published *Nuova Scientia*, in which he discussed the trajectory of a cannon ball. According to his theory, such a trajectory consisted of three parts. First the ball moves along a straight line, but as the speed decreases it enters into a circular path, until it reaches the vertical and falls straight down. Tartaglia succeeded in proving that all projectiles fired at the same angle follow similar trajectories.

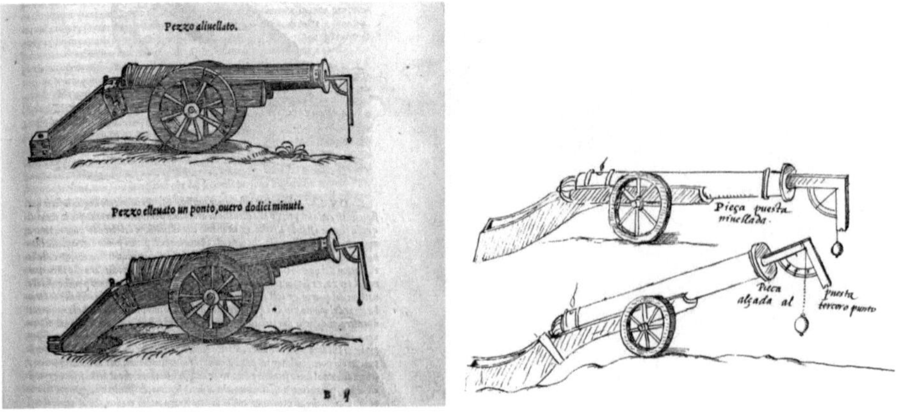

Fig. 11.3 The aiming of a cannon with a gunner's quadrant in Tartaglia and Coignet (1618b) (EHC 520935)

He also demonstrated that a maximum range was obtained with a firing angle at 45° (Fig. 11.3). His theory would be applied by artillerymen for the next a hundred and fifty years.[5]

Another important contribution to the theory of ballistics came from Luis Collado (1606), who had conducted experiments with a falconet, i.e. a 3.5 pound gun, for six subdivisions of the quadrant (7.5°, 15°, ...) and found 594, 794, 954, 1010, 1040 and 1053 paces as ranges. The minimum range for $\alpha = 0°$ was 268 paces.[6] In 1586, he published an artillery textbook in Italian, which he subsequently translated and published in Spanish in 1592 (*Prática Manual de Artillería*, Milan).

In his book *Artillerie*,[7] Diego Ufano, a former captain at the Antwerp citadel, was the first to try and propose a mathematical model for the trajectory of the bullet. Ufano proposed the following formula for calculating the range:

$$S = \frac{200 + \sum_{i=1}^{\alpha}(45°\text{-i})}{200} S_{min}$$

where α is the elevation and S_{min} the minimum range. This formula is more or less in agreement with his empirically (?) found values for a gun with $S_{min} = 200$.

---

[5]Tartaglia must have been aware that the first part of the trajectory was not really rectilinear, as he himself argues that the weight of the ball will cause it to fall and therefore deflect from the ideal path. It is ironic that Tartaglia's theory withstood practical tests, while Galileo's was disproven. One of the reasons is air resistance ("friction"), which, although absent from Galileo's theory, impacts significantly on a ball's theoretical trajectory. See Hall (1969, pp. 38–42).

[6]Hall (1969, p. 46).

[7]Ufano (1614, p. 134), also Hall (1969, pp. 46–47).

In Coignet's *El Uso* (1618), we find Collado's range values. According to Coignet a *tiro muerto* (dead shot), for which the bullet hits the objective at the end of the circular part of the trajectory, is attained at an elevation of 30°. Coignet also gives other range values for different types of cannons. In 1601, Augustin de Mexia, commander of the Antwerp citadel,[8] and his quartermaster Jean Monroy had done some experiments with a 24 pound cannon to establish the maximum range. It is highly unlikely that Coignet was unaware of these experiments and he may even have attended some. In his *Harmonie universelle* (1637–1638),[9] the untiring letter-writing mathematician, Father Mersenne, gives a table with ranges for different elevations and calipers. These are the same figures as those described in Coignet's *El uso*, which explains why Mersenne attributes them to him.

11.3 Fortifications at Antwerp

In 1542, the siege by the rebel army led by Maarten van Rossum made it very clear that Antwerp's defences were hopelessly outdated. Advice was sought from Donati Buoni di Pelzuolli, an Italian engineer, who proposed the construction of bastionized city walls. The realization of his plans would bring the city to the brink of bankruptcy by 1547.[10]

Already in 1565, Granvelle had decided to follow the example of Ghent and Cambrai, and to construct citadels in the principal Netherlandish towns (Antwerp, Valenciennes, 's Hertogenbosch, ...) to keep the cities subjugated and as a safeguard against revolt. Francesco de Marchi was charged with finding a suitable location to erect the Antwerp citadel.[11] His proposal was for it to be built to the south of the city, near the abbey of St Michael's. The Iconoclastic Revolt persuaded Alba to push through these plans as a matter of urgency. The Italian architect Francesco Paccioto of Urbino (1521–1591)[12] was to supervise the construction work, which began on

[8]Genard (1877b, p. 296).

[9]Mersenne (1636–1637, pp. 39–40).

[10]Soly (1977), also Bertels and Martens (2006).

[11]Soly (1976, p. 550).

[12]Duffy (1996, p. 68), cites a letter from Francesco Paccioto to the Count of Urbino, in which he boasts of finishing the work within 4 months! Soly (1976, p. 552), has shown that the accounts mention not Francesco, but Bernardino Paccioti as the chief engineer during the years 1568–1571. Couvreur (1985) argues that the correct name is Bernardino *Faccioto*, the author of *Nova Maniera de Fortificatione* (1570). This is confirmed by van den Heuvel (1991, p. 153). Be that as it may, clearly Francesco Paccioto left the city in March 1568 because he was not getting paid. Alba, in his letters to Philip II, mentions "the fool Paccioto ... who is returning to Italy because his wife and children are hungry." (Duque de Alba 1952 2, nr. 693, pp. 17–20, here p. 19). On the accomplishments of Paccioto, see Kubler (1964) and van den Heuvel (1991, pp. 155–156). Paccioto was a tutor to the young Alessandro Farnese (then 8 years old) and taught him arithmetic. In Italy, he had built the fortresses in Montecchio, Corregio, Guastalla and Scandiano.

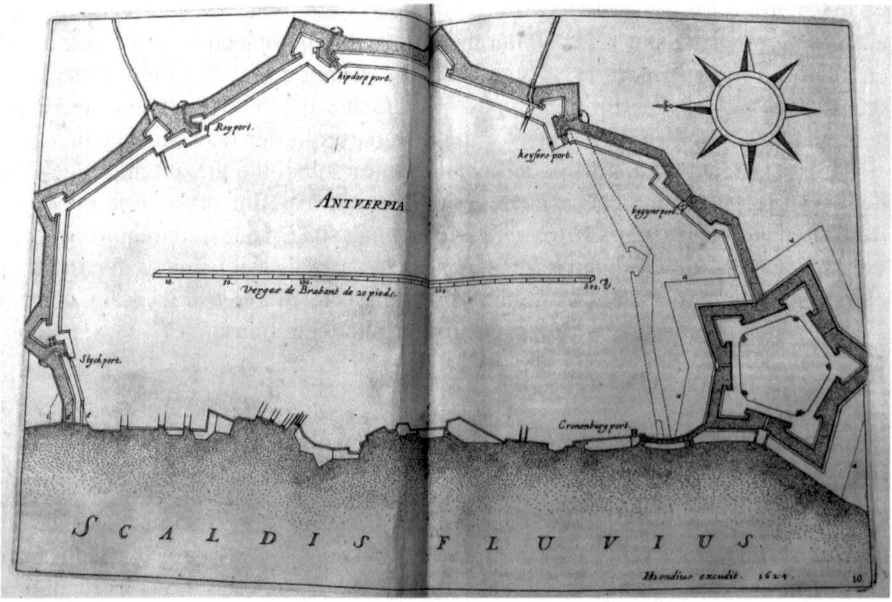

Fig. 11.4 The Antwerp fortifications in Hondius (1624) (EHC G 10414:a)

1 November 1567 and subsequently progressed at a staggering pace, so that the building was completed in just a couple of years (Figs. 11.4 and 11.5).

The choice for the site was a gamble, as the Spanish had no idea from which direction either reinforcements or attacks were to be expected. As a matter of fact, from a defensive point of view, it turned out to be the wrong bet. With insurgents commanding the terrain north of Antwerp, the citadel was not strategically located. When Farnese launched his counterattack from the south, the structure actually stood in the way of the Spanish troops! It did however fulfil its other purpose of keeping the people of Antwerp down.[13]

Throughout this period, Peter Frans was the city architect. Appointed in 1540, he remained in office until 1583. Among his apprentices were well-known names such as Adriaan Bosch, Abraham Andriessen, Hans van Schille, Hans Vredeman de Vries and Daniel Speckle.[14] Actually, they were quite a mixed bunch, as becomes clear from their biographical details.

Little is known about Hans van Schille. He became a master of the Guild of St Luke in 1521 (or 1532?).[15] Boys were apprenticed to him to become painters. He

[13]Duffy (1996, p. 68) and van den Heuvel (1991, pp. 107–119).

[14]Müller (1959, p. 469). On these architects, see also van den Heuvel (1991), Westra (1992).

[15]Rombouts and van Lerius (1874, I, p. 98 and 118).

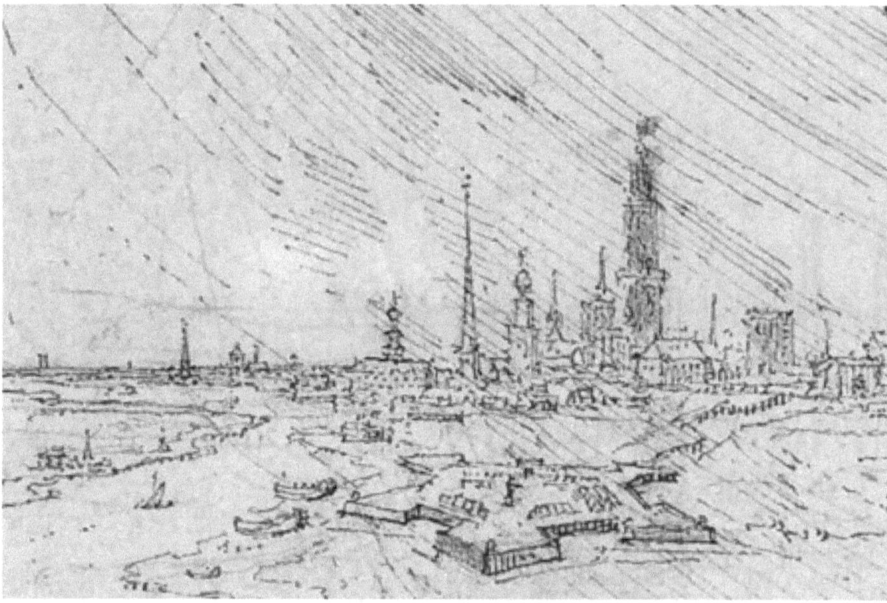

Fig. 11.5 View of Antwerp and the Scheldt from the south, with Alba's citadel in the foreground. Detail of van Schille's entry in Vivianus' *Album Amicorum* (Vivianus (1571–), f52*r*, Koninklijke Bibliotheek Den Haag 74F19)

may also have worked as an *erfscheider* (surveyor).[16] In 1567, he was forced to flee after it became known that he had illustrated some publications denouncing the Inquisition.[17] In the 1570s, he was employed by the States-General, while in 1577 he became "superintendent and chief engineer" of the city fortifications. He would remain in this capacity after the Reconciliation.

Abraham Andriessen began his career as a contractor. Together with Dierick van Molle he was charged with the job of demolishing part of the citadel. In 1577 he, together with Peter Frans, became a superintendent of the fortification works and was appointed as deputy to Hans van Schille.[18] The next year he was appointed by the States-General as superintendent of the fortifications of Dendermonde (Fig. 11.6) and Herentals, for which he took the oath as an engineer.[19] He was engaged in several construction works for both the city of Antwerp and the States-

[16](Wauwermans, 1894, p. 33).

[17]Marnef (1992, pp. 222 and 228).

[18]SAA Pk2442, dd. 16 November 1576 and Pk552, f26*v*.

[19]Westra (1992, p. 24).

Fig. 11.6 Dendermonde in the sixteenth century (MPM-SPk B 11)

General.[20] He was ennobled in recognition of his services. Andriessen died a rich man from dysentery in 1604 on a voyage to Calais.[21]

Hans Vredeman de Vries is better known for his work on perspective and as a painter, but he was also employed as an architect. He was made a citizen of Antwerp in 1548.[22] In 1570, he got into trouble with the authorities and, as so many others, fled to Germany, where he settled in Aachen. In 1572 he wanted to return and Marten de Vos, the Dean of the Guild of St Luke, and Peter Baltens vouched for him, knowing him "as an honest and devout man, of good name and fame".[23] In the years that followed, he worked as a painter, a designer and an engineer/architect (Fig. 11.7 for one of his architectural drawings).[24] He was involved in draining the polders, casting up dikes and improving the defences outside the city.[25] Together with Abraham Andriessen, he drew up plans for defensive walls that would also incorporate the suburbs of Berchem, Dam and Kiel, but these would never materialize. In 1586, he travelled to Frankfurt, the beginning of a peripatetic life that took him to Brunswick, Hamburg, Danzig and Prague (where he resided at the court of Rudolf II[26]). According to Carel van Mander, the biographer of sixteenth century painters, it was Gillis Coignet who, around 1600, talked him into coming to Amsterdam. Vredeman is an important figure in the history of art. His style

[20]SAA Pk554, f174*r*; Pk658, f283*v*; Pk662, f68*v*; GA4833, f21*r*; R1776, f74*r*.

[21]SAA WK281, f321ff., Andriessen was given medical care by Lazare Marcquis and Jacob Coignet, who would also embalm his body. Genard (1886, p. 86).

[22]SAA V146, dd. 5 September 1548.

[23]SAA Cert33, f351*v*.

[24]SAA R24, f249*r*; Pk 558, f12*v*.

[25]SAA R1778, f74*r*.

[26]Fucikova (1988a, p. 189) and Fucikova (1988b, p. 60).

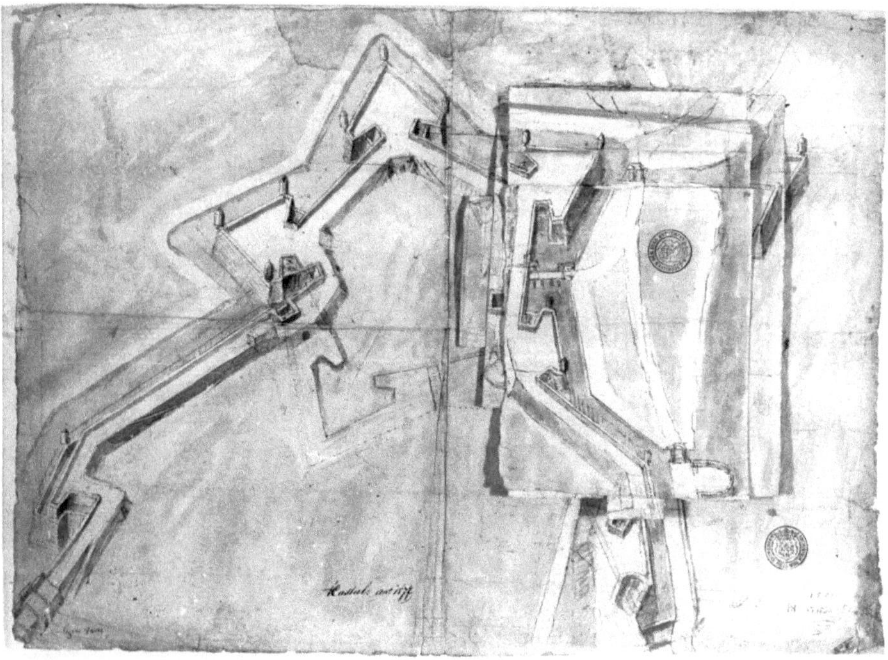

Fig. 11.7 Plans for the incorporation of the citadel into the city walls by Vredeman de Vries (SAA Icon 26/4)

of ornamentation would influence subsequent artists and mathematicians in the Netherlands, northern Germany and the Baltic.

Remarkably, the name Coignet is not mentioned in documents from this period. Michiel almost seems to have vanished from the earth: there is no mention of him other than in a couple of administrative documents. With his knowledge, one would expect him to have been involved in some capacity or other in the defence of the city, but this is not the case. On the other hand, we do know that, on 8 October 1585, he was admitted to the *kolveniersgilde*, one of the six armed guilds that had been raised to defend the city alongside the Spanish troops. Coignet's name is first mentioned in connection with the city's fortifications in 1592, when he was requested by colonel Mondragon to draw up a plan of the fort (i.e. the citadel to the south of the city).[27] Later, in 1607, he would also design several forts at the request of the City.[28]

After 1596, when he came into the service of the Archdukes, Coignet supervised several projects around Antwerp and also attended at sieges. His advice was sought during the sieges of Hulst and Ostend.[29] It is not entirely clear in what capacity

[27]SAA R32, f285*v*.

[28]SAA R47, f326*r*.

[29]Pinchart (1860, p. 294); ARA Raad van State en Audiëntie 2654, dd. 31 July 1624.

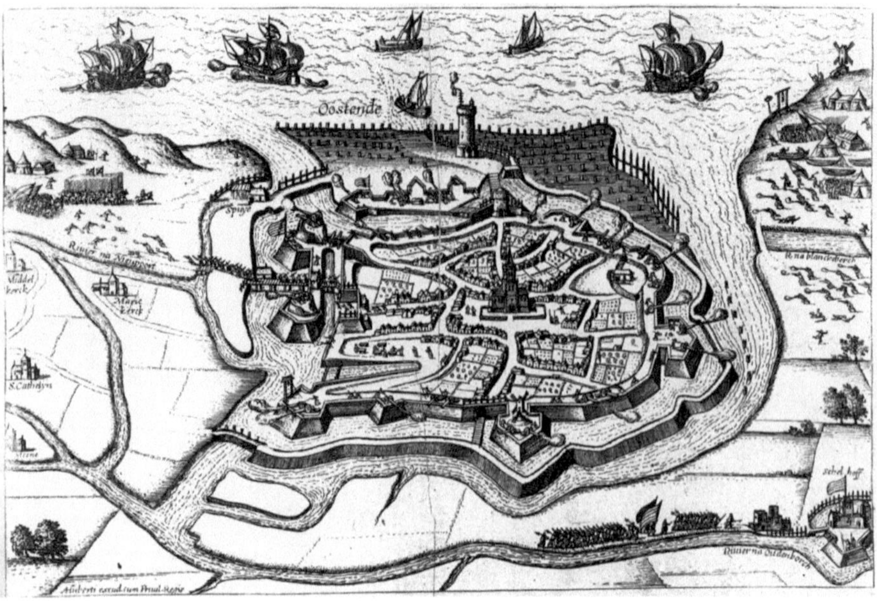

Fig. 11.8 The siege of Ostend. Note that the city could be supplied by sea (SAA 12#338)

he acted, but given his interests it is quite likely that he made calculations for the artillery and he may also have advised on the strength of the fortifications.

The siege of Hulst began on 8 July 1596. Initially, the defences held out well. But the Spanish bombardment of the city was relentless, with 1,000–1,500 shells fired daily. Many mines were detonated to try and breach the walls. But still Hulst held out and the commander in chief, Count Solms, called on the Dutch States to send an engineer to help close the breaches. On 16 August, no fewer than 3,200 shells were fired at the town, but still Albert did not attack. Then, around the time that Albert began to wonder whether the siege could ever be successful, Hulst unexpectedly surrendered. [30]

Ostend was a completely different story: the town could could be supplied by sea and Stevin's system of sluices to control the water level in the surrounding terrain was used as a means of defence (Fig. 11.8).[31] The siege of Ostend was the ultimate engineer's war. Hence it is no surprise to find Coignet fully involved. This,

---

[30] See *s.n. (1964)*.

[31] Thomas (2004c) in particular Thomas (2004b). On Stevin's ideas, of which Ostend was an example avant la lettre, see Dijksterhuis (1970, pp. 109–110). Already in the 1580s, Thomas Digges reported to Burghley and Walsingham that the city "(is) strong by situation because by means of sluices the country can be drowned or drained at pleasure ... the access being upon strait banks easily fortifiable". He went on to say that the harbour entrance was difficult to defend, but that this could be remedied by moving it to the centre of the city (Calendar of State Papers, Foreign (1585–1586), cited in Webb (1945, pp. 395–396)).

Fig. 11.9 Equestrian portrait of Ambrogio Spinola (Verhoeven 1621, EHC B 26881)

the longest siege of the 80 Years' War, began in July 1601. The city of Ostend was the last bridgehead of the Dutch States in the Southern Netherlands.

The Spanish army first attempted to drive the defenders back within the city walls. After a heavy bombardment during the night of 6–7 January 1602, the assailants launched an offensive across open terrain. It turned out to be a disaster, and they would not attack again until the Summer of 1603, as the siege and bombardment continued.

Towards the end of the Summer of 1603, Ambrogio Spinola (Fig. 11.9) became the new commander of the Spanish Army. Spinola had read extensively about war, but he was inexperienced. This was his chance to demonstrate his worth as a military commander. He successfully launched an attack and pursued the defenders beyond the city walls, so that his troops could now besiege the inner city. The defenders soon had to withdraw to a small fortified part of town. Moreover, heavy storms made it impossible for supply ships to reach them. On 20 September 1604, Ostend finally fell after a siege that is said to have cost the Spanish some 40,000 able men. After

Fig. 11.10 Fort Liefkenshoek now

the Sack of Ostend, Spinola launched a successful offensive against the troops of Prince Maurice in the North.[32]

Around 1600, Coignet took part in the inspection of the fortifications at *Vlaams hoofd* in the Waasland on the left bank of the Scheldt.[33] These fortifications were crucially important. For the Spanish, it was paramount that the insurgents should be driven from as many forts along the banks of the Scheldt as possible, as these allowed them to control traffic on the river. Two forts, Lillo and Liefkenshoek (Fig. 11.10), were of particular value in this respect. Despite the best efforts of the Spanish, the Dutch held these forts until 1638. In 1599, Prince Maurice mounted an attack from Liefkenshoek, which brought him to the gates of Fort Vlaams Hoofd.

In 1608, Coignet, together with Matthias van Herle, drew up the plans for the fortifications near the Abbey of St Michael's.[34] Other maps attributed to Coignet are believed to date from around 1614, when he was inspecting the dredging works on the city canals. He complained that one of the contractors, Martin Anckels, had dug too narrowly and that he had literally cut corners to minimize his work. Coignet noticed that "many poles were displaced during the night time", so that he decided to inspect the works repeatedly.[35]

From 1616 onwards, his name appears in the city accounts in connection with work on the Antwerp fortifications. In most cases, it concerns the granting of permission to the comptroller to pay for certain services.[36] As an engineer, he was

[32]On the siege of Ostend, see Thomas (2004c).

[33]ARBW, Land van Waes 137, f36*v*–37*v*

[34]SAA Icon Kaarten III/3. Van Herle was one of the official city surveyors (SAA R33, f131*r*).

[35]SAA Pk712, f48*r*–49*v*.

[36]ARA Rekenkamer 26279, f15*v*–16*r*, f18*v* ff.; 26280, passim.

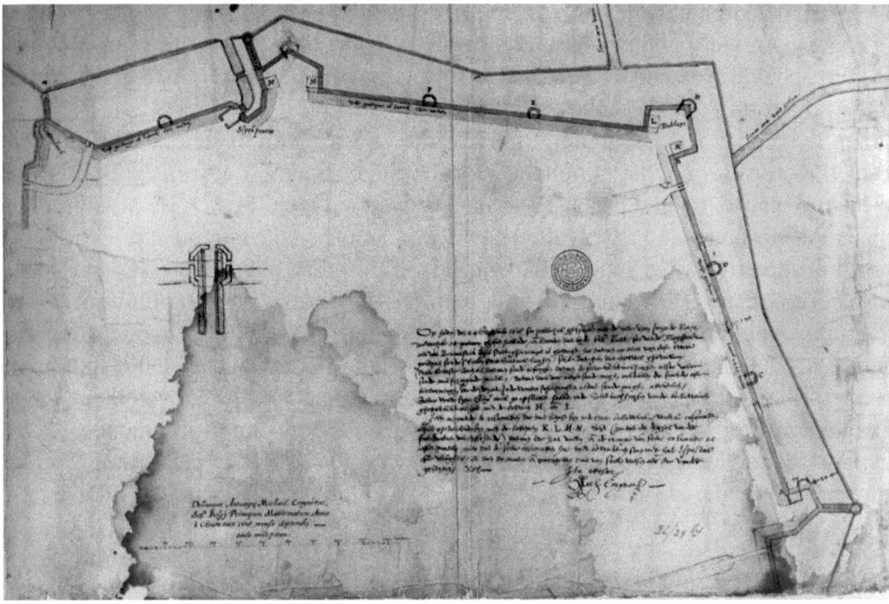

Fig. 11.11 Plans for the city fortifications by Michiel Coignet (ca. 1618, SAA Icon 26/4)

involved in the construction of Fort Isabella[37] on the left bank of the Scheldt,[38] the reinforcements at the so-called *Vlaams Hoofd,* also on the left bank,[39] and possibly the fortifications at the nearby town of Lier.[40]

In 1621, Coignet was paid 40 gl. for drawing five maps of a canal for the Kleine Nete. The purpose of these maps, which are now lost, is not clear. Cutting a canal between the Kleine Nete and the Grote Nete (upstream from their confluence in Lier) had always been an ambition of Antwerp.[41] The canal would serve two purposes. First and foremost, it would benefit trade. Shipping on the Nete was hampered by the fact that it had to pass through Lier. Digging a canal would make it possible to navigate directly from Herentals, up river, to Antwerp. Second, it would guarantee an additional supply of fresh water, which always seems to have been a problem in Antwerp. Plans for such a canal date back to the beginning of the fifteenth century, but they only materialized at the beginning of the eighteenth century.[42]

---

[37] Lombaerde (2009, p. 54).

[38] "... when he was making the new fort north of the Vlaams Hoofd ...", SAA Pk712, f48*r*–49*v*.

[39] ARA Rekenkamer 26280, passim. Lombaerde (2009, pp. 54–55).

[40] ARA Rekenkamer 26280, f35*r*.

[41] SAA R63, f359*v*.

[42] Prims (1933).

In 1620, Coignet was appointed as superintendent of the works on the fortifications near Antwerp, which were intended to provide protection against naval attack via the Scheldt. Any enemy fleet sailing upstream would be exposed to gunfire from one of the forts.

In September 1618, Coignet discussed the construction of sentinel houses with Don Iñigo de Borgia, the commander of the Spanish garrison.[43] Possibly this was the same plan as he had proposed to the city council (Fig. 11.11). He also drew up plans for the canals in the so-called new town, an urban development project to the north of the city. After visiting the site, he proposed to build two casemattes near St George's Gate. He visited the area with Burgomaster Dasse to point out where the guns could be positioned to protect the Scheldt.[44] No fee was agreed for these services, but Coignet did make it known to the City Council that he would appreciate some form of remuneration. A source mentions that in other "cities and places where he had delivered such services he always asked a modest sum in recompense",[45] so apparently he worked in quite a few other places, too. The city paid him 250 gl. [46] for services rendered since 1614.[47]

[43]SAA Icon Kaarten 26/24bis.

[44]SAA Pk712, f48*v*–49*v*.

[45]SAA Pk712, f48*v*–49*v*.

[46]SAA Pk712, 49*v*. de Burbure (s.d.), I.265.

[47]SAA R60, f334*r*; R61, f334*r*.

Chapter 12
Conclusion

Although the main focus of this study is on Michiel Coignet, the reader will no doubt have noticed that another character also features prominently: Gemma Frisius. The numerous connections between Michiel Coignet and Gemma Frisius are illustrative of the influence of the latter on the mathematics scene in the Netherlands of the sixteenth century. Gillis Coignet sr. may have cooperated closely with Frisius, as some of his instruments bear a close resemblance to those proposed by the latter. Moreover, a number Frisius's students are within Coignet's circle of friends, e.g. Mercator or Gualterus Arsenius. Ferdinand Arsenius, a member of this family, worked as an engraver in Coignet's workshop.

We may regard Michiel Coignet as the last representative of Frisius's school. At some point, he seems to have continued along the path indicated by Frisius. When calculating the length of a loxodrome or publishing a Mercator chart, he must have been in clear sight of the watershed between the old and the new. But like Stevin, he never quite reached it. Galileo and Kepler would lead the way to the paradise of infinitesimals. The Jesuits were other guides on that road. They made their calculations using Coignet sectors. In a sense, then, Coignet was "instrumental" in making possible the change that was taking place.

Coignet also seems to have been a typical example of a -high end- mathematician of the commercial centres of the Low Countries and indeed Western Europe. Strongly rooted in practical mathematics, they sought to solve the problems that the new age posed. Starting out as a promising mathematician, after 1585 Coignet increasingly turned his attention to engineering, no longer contributing original work but focusing instead on the practical application of mathematical knowledge.

Coignet was familiar with the books of his contemporaries and cites them extensively in his own work. The panorama suggested by these references and the value of his library is a formidable one, even taking into consideration that he lived in the then capital of printing.

All the fields Coignet was involved in – wine gauging, surveying, fortification and ballistics – involved remarkable mathematical developments and techniques, supported by new, quite ingenious mathematical instruments. Coignet was well

A. Meskens, *Practical Mathematics in a Commercial Metropolis: Mathematical Life in Late 16th Century Antwerp*, Archimedes 31, DOI 10.1007/978-94-007-5721-9_12,

© Springer Science+Business Media Dordrecht 2013

versed in all of these fields and he even contributed to them, albeit within the conceptual framework of sixteenth-century mathematics. In his mathematical work, Coignet was no revolutionary innovator, yet his influence was further reaching than has hitherto been assumed. His sectors were widely used and enabled the Jesuit mathematicians to perform their calculations. Manuscripts on the sector appeared in every Western European language, except in Coignet's native tongue.

Coignet appears always to have remained a modest man and to have avoided the limelight. Yet he must have enjoyed a considerable reputation among his contemporaries and is often named as one of the great mathematicians of his time. The Jesuit Carolus Scribani, Rector of the Jesuit college in Antwerp, in a letter to Erycius Puteanus of 1625, refers to Coignet and Franciscus d'Aguilon s.j. as the greatest mathematicians of the Netherlands.[1] Unfortunately, there are no extant portraits of Coignet. None of the surviving copies of his books has a printed portrait of the author, as so many other books of the time did. In his biography of men from the Low Countries, *Athenea Belgica*, Franciscus Sweertius claimed that Coignet was held in high esteem by the Archduke as well as by the Marquis Spinola. In his necrology of Archduke Albert, *De vita Alberti pii* (1622), Aubertus Miraeus writes that the Archduke held both Coignet and the painter Wenceslas Cobergher in high regard as mathematicians. Apparently, Albert took Coignet as a mathematics teacher[2]; according to Bruslé de Montpleinchamps, he actually taught Albert all he knew about mathematics.[3]

As a court mathematician in a country ravaged by war, he had to contribute to the defences. Even in his old age and nearly up to his death, he was involved with the fortifications.

Thus ended a life that had begun so promisingly in a time of economic prosperity and commercial expansion, but which had been turned sour by the ravages of war. His fate stands in sharp contrast to that of his counterparts who chose to leave the south, including Simon Stevin.

Mathematics, which at one time flourished, all but disappeared in Antwerp. The once bustling port, where printers published books for all possible audiences, became a mere shadow of its former self. Antwerp gave Amsterdam "its tired, its poor, its huddled masses yearning to breathe free. It sent these, the homeless, tempest-tost masses" north to begin new lives. Their pursuit of happiness would build Holland an empire. Amsterdam would succeed Antwerp as the leading port of the Low Countries. More often than not, the ventures that made it prosper had been established by emigrants from Antwerp. The mathematical thread that was cut in Antwerp was soon picked up with great success in Amsterdam.

[1] Brouwers (1972, pp. 150–154, here p. 152).

[2] de Maeyer (1955, pp. 366–367).

[3] de Montpleinchamp (1693, p. 372).

Appendix

Unless otherwise indicated, all references are to archival source in Stadsarchief Antwerpen (Antwerp City Archives). Hence the prefix SAA is omitted.

A.1 Emigrated Antwerp Arithmetic Teachers

See Briels (1972b; 1972c, (B14), 1973, (B15)) and van Roosbroeck (1968) (R).

Reference	Name	Emigrated to
B14 283	Jaspar Beck	Den Bosch (±1577); Delft (1589)
B14 295	Bartholomeus Cloot	Delft (1574)
B14 298	Jan Coutereels	Delft (1594); Middelburg (mei 1594); Arnemuiden (1613); Middelburg (1622)[a]
B14 291	François Jansz de Bruyne	Amsterdam (1589)
B14 290	Anthony van den Broecke	Rotterdam (?); Delft (1591)
	Hendrik Dens	Cologne (1592)[b]
	Christian de Riche	England (1566)[c]
R196	Jeronimo de Roye (?)	Frankenthal (na 1581)
R183	Jacques de Vos	Frankfurt (1572)
R307	Paschier Goossens	Hamburg (1594)[d]
R138	Arnold Hesius	Cologne (1568)
B15 118	Peter Heyns	Delft (±1585); Frankfurt (1587)[e]; Stade (1589); Haarlem (1594)
R138 255 305	Peter Heyns	Cologne (1568); Hamburg (1591)
B15 128	François Loockmans	daughter marries 1591 at Middelburg
B15 143	Hans Peeters	Gouda (1590)

(continued)

A. Meskens, *Practical Mathematics in a Commercial Metropolis: Mathematical Life in Late 16th Century Antwerp*, Archimedes 31, DOI 10.1007/978-94-007-5721-9,
© Springer Science+Business Media Dordrecht 2013

(continued)

Reference	Name	Emigrated to
B15 140	Kerstiaan Offermans	Haarlem (1586)
B15 269	Michiel Six	Middelburg (1585)
B15 270	Anthony Smyters	Stade (1590); Amsterdam (1600)
B15 273	Bernard Stockmans	Dordrecht (1589); back to Antwerp (±1595)
	Hans van der Veken	Rotterdam (?)[f]
B15 123	Robrecht van Huesden	Zaltbommel (?); back to Antwerp
B15 131	Aert van Meldert	Rotterdam (1585)
B15 263	Felix van Sambeke	Delft (1586)
B15 266	Francois van Schoonhoven	Leiden (1589); Amsterdam (1607)
B15 281	Jacques Vercammen	Flushing (1590); Den Briel (1599)
	Isaac Vullevens	Dordrecht[g]

[a]A.P. BUYS, personal communication
[b]Meeus (1988–1989, p. 607)
[c]Page (1893, p. 75)
[d]also: Brulez (1960, p. 291)
[e]SAA SR399, f10v–12r. He stayed here at least until the beginning of 1589, when his son Zacharias stayed with him. Zacharias was sent to Frankfut for the *Buchmesse* by his patron Jan Moretus (Rooses and Denucé 1968). See also Meeus (1988–1989)
[f]Brulez (1960, p. 295)
[g]De Groote (1967a) and De Groote (1968d, p. 188)

A.2 Schoolmistresses 1585–1620 and Schoolmasters 1600–1620

List of schoolmistresses and schoolmasters who taught arithmetic in respectively the periods 1585–1620 and 1600–1620. For the sixteenth-century arithmetic teachers, we refer to the studies by H.L.V. De Groote.

		A4529	
1585	Jacomijnken Willem	f103*v*	ciphering
1589	Anna van der Mast	f139*r*	counting with jetons
1590	Maria van Eeckhoudt	f148*r*	arithmetic
1591	Sara Nouts	f148*r*	arithmetic
1592	Ursula de Neve	f170*r*	counting with jetons
	Françoise van Ommeslaghe	f170*r*	counting with jetons
	Maeyken van Eynde	f170*r*	counting with jetons
1593	Catharina Voorlinckx	f174*v*	counting with jetons
	Maria van Eynde	f185*r*	counting with jetons
1594	Ursula de Neve[a]	f186*r*	counting with jetons
	Elizabeth van de Putte	f186*r*	counting with jetons

(continued)

(continued)

1595	Elizabeth Jacobs	f196*r*	rekenen
1596	Soetkin Cornelis	f208*r*	ciphering up to the rule of three for integer numbers
	dochter van Merten Nijs	f208*v*	counting with jetons
1597	Anna Borrekens	f209*v*	arithmetic
1598	Janneken de Vos	f225*v*	counting with jetons
	Susanna Adriani	f234*r*	counting with jetons
1599	Elizabeth van Assche	f244*v*	continues her deceased husband's school
	Anna Foots	f245*r*	counting with jetons
1600	Maria de Wijse	f246*r*	counting with jetons
		A4530	
	Jacques Pigaiche	f3*r*	ciphering
	Maria vanden Omslach	f3*r*	counting with jetons
1601	Hans Beucker	f13*r*	ciphering up to the rule of three
	Diricx van Merxem	f13*v*	ciphering
	Isabelle Gramaye[b,c]	f13*v*	ciphering up to the rule of three
	Clara Vergods	f14*r*	counting with jetons
	Adriaan Ducentius	f14*v*	ciphering and accounting
	Jan Ruraet	f15*r*	ciphering up to the four species
1602	Anna Eedtvelt	f27*r*	arithmetic
1604	Jacques de Menil	f40*r*	ciphering and accounting
	Robbrecht Janssens	f41*r*	ciphering up to the four species
1605	Gillis Vencker	f66*r*	ciphering up to the rule of three
1607	Nicolaes Claessens	f85*v*	ciphering
1610	Lenaert de Raeymaker	f107*r*	ciphering and accounting
	Jacques Hoydock	f107*v*	ciphering and accounting
1611	Philips Franchoys	f115*r*	arithmetic
1612	Margrita van Wijck	f122*r*	arithmetic
1613	Anna Neirejans	f130*r*	counting with jetons
1614	Jan Vermeiren	f137*r*	ciphering and accounting
	Clara Hayens	f137*r*	arithmetic
		A4530	
1615	Cornelia vande Castele	f144*r*	counting with jetons
1616	Joris Bertice	f151*r*	ciphering and accounting
1619	Anthony Vranckx	f173*r*	ciphering and accounting
1620	Peter van Leuven	f188*r*	ciphering up to the rule of three

[a]This is the same Ursula de Neve as mentioned in 1592. At that time she was admitted as an assistant teacher, in 1594 she opened her own school

[b]This is the same Ursula de Neve as mentioned in 1592. At that time she was admitted as an assistant teacher, in 1594 she opened her own school

[c]Daughter of Cornelis I Gramaye, tenant of the wine excise

A.3 Religious Convictions of the Arithmetic Teachers

A.3.1 According to SAA A4830(1) Lijsten Borgerlijcke Wacht

C = Catholic, L = Lutheran, R = Calvinist of Anabaptist.

Mr. Peter[a]	C	417*v*	Mr. Dierick	C	417*v*
Mr. Jan de schoolmr[b]	C	4*v*			
Carel Baillule	C	398*v*	Jan Borkens	C	380r[c]
Michiel Coignet	R	52*v*	Symon Colyn	R	379*v*
Aert de Cordes	R	169*r*	Jan de la Court	K	295*r*[a]
Jaspar de Craeyere	K	377*r*[e]	Mr. Peeter Goossens	R	423*v*
Hans Hermans	C	227*v*	Mr. Arnoult Hesius	L	385*v*
Hans Hesius	L	385*r*	Peter Lottyn	R	21*v*
Claude Luython	C	194*v*	Noel Morel	C	64*v*[d]
Xpiaene Offermans	R	248*r*	Jan Rademaker	R	178*v* of 181*v*
(Peeter) Hans Symons	L	432*v*	Mr. Jan van Brande	C	447*v*
Jan van den Bossche	C	401*v*	Peter van Schelle	R	49*r*
Mr. Adriaan Verdonck	C	423*v*			
Rombout Verdonck	C	55*r*	Mr Willem his assistant teacher	C	55*r*

[a]Little is known about Mr. Peter and Mr. Dierick. The only information provided is that they were assistant-teachers
[b]Possibly Jan van Swolle, who lived at Oude Korenmarkt
[c]Centurion. Also Pk2566, f26*r*
[d]Decurion
[e]Centurion. Also SAA Pk2566, f26r

A.3.2 According to A4550 Schoolmeestersgilde

The following names were marked with an O in a list drawn up in 1576. This list does not correspond entirely with the registers of the Guild. The O identified the individuals concerned as Protestants.

Arnout (=Aert) de Cordes	Felix (Sambeke?)
Franchois Schoonhoven	Hans (Jan) van den Bossche
Jacob Crekels	Jacob Arondeaulx
Jacob van den Brande	Jacob metten gelde
Jacob de la Court	Jacob Beuckelaar
Merten van Dick	Niclaes Bormans
Robrecht Huesen	Simon Moons
Wouter de Costere	Ghijsbrecht Gheyssels

These schoolmasters signed the accounts of Dean Jan Borrekens. Peter Heyns had transferred the accounts of the Guild to Jan Borrekens when he left the city. On 1 April 1586, they were transferred in the usual way to the new Deans. It can be surmised that these signatories were Catholic, because by then the Guilds had been religiously cleansed.

S. Moons	P. de Ayala
B. Van Vlierden	Fabius
R. Verdonck	

Left Antwerp immediately after the Reconciliation:

Peter Heyns, Michiel Six, Jacob van Houthuys

A.4 Residence of the Schoolmasters

This table contains all known residences of schoolmasters. The column R denotes the rent charge value (*rentewaarde*) of the house, an O in column E denotes an owner, a T a tenant. The rent charge value was the basis for taxation. It was theoretically equal to the value of the rent charge (see footnote 7 on page 10) which could be sold on the house. Since this last value was 1/16th of the value of the house, its theoretical value can be calculated. The rent charge value thus becomes a basis for comparison for real estate prices.

Name	Name house	Street	R	E	Source
Karel Bailleul	In de Gulden Hant	Peestraat		O	V1407 142*v*
Assuerus Boon	't Vliegent Peert	Venusstraat			A4529 47*v*
		Wolstraat	292	H	R2241 9*r*
Jan Borrekens	In Apollo	Lange Nieuwstraat	150	O	R2224 5*v*
					R2234 9*v*
					R2321 19*v*
					R2248 21*v*
Michiel Coignet	Eekhoorntje	Braderijstraat[a]	100	T	SR371 186*v*
					R2237 48*r*
					R2314 47*r*
					R2333 45*v*
					R2350 48*r*
Prosper d'Albora	In Adam en Eva	Lange Meystraat	20	T	R2320 78*r*

[a]On 25 September 1582 Michiel Coignet and Maria vanden Eynde took over the rent of the house (155 gl./year) from Anna van Aelst, wife of Tristram Verhoeven. The house was owned by Gaspar Verryt, who had bought it in 1580 (SAA Pk2257, f280r). It was rented for a period of 6 years beginning in 1581. Some ledgers wrongly mention Melchior Quinget

Name	Name house	Street	R	E	Source
Christoffel Daneau		Cathlijnevest			V1404 94*r*
Arnout (=Aert) de Cordes	Turnhoudt	?	40	T	R2248 41*v*
		Achterstraat	24	T	R2241
Hercules de Cordes[a]	De Penne	Breestraat	45	O	R2226 3*r*
Wouter de Coster	(een love)	Kerckhofstraat			SR276 215*r*
Jan De Crayer	Het Pemelshuys	Lange Nieuwstraat	300	T	R2234 3*r*
					R2321 6*v*
					R2247 8*v*
Adriaan Dens	Inde Poorte	Maelderijstraat[b]	100	T	R2181 131*v*
Hendrik Dens		Zuurstraat[c]	150	T	R2318 92*v*
Jan van den Bossche	Slot van Milanen	Coepoortstraat	80	O	R2210 21*v*
					P2261 312*r*
			80		R2241 22*r*
Jan de Rademaker[d]		Korte Nieuwstraat	80	T	R2318 80*r*
Jan Desbonnets		Kipdorp	110	T	R2232 106*v*
					R2321 100*v*
Jacob de Vos	't Gulden Leeuwke[e]	Clarenstraat	100	T	R2247 40*r*
Arnout Gillis	Craeyenborch	Kipdorp	60	O	R2232 70*v*
					R2247 71*v*
Jan Goddaert[f]		Huidevetterstraat	80	T	R2320 17*r*
Passchier Goossens		Lombaerdevest		T	A4833 292*v*
Peter Goossens		Lange Nieuwstraat	300	T	R2230 5*v*
					R2247 7*v*
Arnout Hesius		Kipdorp	150	O	R2232 118*v*
					R2321 112*r*
Peter Heyns		Cloosterstraat		O	A4833 371*r*
Bart. Huysmans		St Pietersvliet	100	O	R2322 97*v*
					R2249 108*v*
Peter Keppens		Schrijnwerkersplaats	108	T	R2220 19*r*
Hans Lemmens		Keizerstraat	50	T	R2248 90*r*
Francois Loockmans		Moriaen	40	T	R2229 144*v*
					R2322 142*r*
					R2249 157*v*
Peter Lottin	Berrye	Wijngaertbrug		T	A4833 4*r*
		Melkmarkt	150	T	R2213 138*r*
Gabriel Meurier	De Gulden Penne	Koepoortbrug	100	E	R2241 101*r*

[a]de Cordes died before 1585. In 1586, his widow sold the house (Also SAA SR338)
[b]"onder den thoren"
[c]=Sudermanstraat
[d]The names of tenant and owner were struck through on 7 February 1584. The house was sold to a certain Vermijlen, who subsequently lived in it
[e]The house was sold around 1585 to Philips van Mechelen, who succeeded in having the rent value reduced to that of the adjacent houses, 60 gl.
[f]Rents the house together with Sebel (Sybille?) Noblet

Name	Name house	Street	R	E	Source
Kerstiaan Offermans	De Lauwerboom[a]	Augustijnenstraat	100	T	R2225 13*v* R2243 15*v* R2315 13*v* A4833 377*v*
Hans Peeters		Brilstraat	75	T	R2248 105*v*
Cornelis Phalesius		Kipdorp	120	T	R2232 96*v* R2321 91*r*
Andries (van) Roosen		Schuydtstraat	40	T	R2244 115*v*
Michiel Six		Dries	102	T	R2249 171*v* R2229 168*v* R2322 158*r*
Anthony Smyters	Het Calff	Kipdorp	216	E	R2234 69*v* R2321 127*r*
Karel Strytbergher[b]		Oude Lombaertstraat	48	T	R2211 62*v* R2242 63*v*
		St Joriskerkhof	20	E	R2245 105*v*
Melchior van Aelst		Pelgrimstraat	32	T	R2181 102*v*
Jan van den Bossche	Slot van Milanen[c]	Coepoortstraat	80	O	R2210 21*v* P2261 312*r*
			80[d]		R2241 22*r*
Jan van den Brande		Faconsplein	45	T	R2322 142*v*
Anthony van den Broecke		Hoochsetterstraat	108	T	R2321 57*v*
Maarten Vanden Dycke	Inde Losch	Niestrate	60	T	R2229
		Cooperstraat			R2322 186*v*
	De Gulden Bock	Catteveste[e]	80	T	R2318 69*v*
Hans vander Veken	Het Lammeken	Wijngaertbrug		T	A4833 1*v*
Andries van Deurne	In Dansick	Dries	56	T	R2229 156*v* R2249 170*r*
Melchior van Elselaer	De Spiegel[f]	Oude Beurs	200	T	R2317 117*v* A4833 61*v*
	Julius Cesar	Zakstraat	84	T	R2240
Jan van Hemert	De Gulden Passer	?	54	T	R2240 150*v*

[a]When the seventh 100th penny tax was collected in 1584, it was noted that Peter Heyns was the owner-occupier of *De Lauwerboom*, but on 17 March 1584 it was added that Kerstiaan Offermans was a tenant. According to Prims (1929), Heyns was a tenant himself from 1571 to 1585. The archives prove him wrong: Heyns was in fact the owner of the house. Also Meskens (1993a)

[b]"owns the school of St George"

[c]This house also had a "voorhuis" which was rented by Jan van den Bossche. The "voorhuis" had an estimated rent value of 60 gl.

[d]Written in the margin: " verclaert maer behoort 40 guld te stane" (declares that it should read 40 gl.)

[e]Catteveste = St-Kathelijnevest between Kipdorp and Lange Nieuwstraat. The name Maarten Vanden Dycke is struck through and replaced with Jacob Stoelman, suggesting the former moved to Koepoortbrug at the time of the fourth collection of the fifth penny tax (A4550)

[f]Melchior van Elselaer lived in a part of the house *De Spiegel*. The whole building had a rent value of 600 gl.; it was shared by three occupants

Name	Name house	Street	R	E	Source
Robrecht van Heusden jr.	Castel van Fonteynen	Lombaardevest			P1510 35*r*
Jacques van Houthuysen	De Roosenboom	Reynderstraat	72	O	R2221 50*v*
Aert van Meldert	St Jacob	Lange Nieuwstraat	80	T	R2230 8*v*
Felix van Sambeke	Int Poortken	Moriaen	120	T	R2249 161*v* R2322 148*r*
Philips van Woelput		Corte Clarenstraat	61	T	R2232 42*r* R2321 41*r* R2247 43*r*
(voor) Merten Vermander		In Predikherenconvent		T	A4833 174*r*
Merten Wouters		Lange Nieuwstraat	36	T	R2246 80*v*

A.5 Revenue from Wine Gauging

Naam	Period of activity of the wine gaugers
Jan Morchoven [a]	before 1550–before 1560
Bernaert Doudens	before 1550–after 1561
Peter van Aelst	1551–1555?
Arnolt vanden Borne	± 1560 ?
Ryckart de Raeymaker	about 1564 ?
Aert de wijnroeier	about 1567
Peter de Cock	before 1572
Carel Blommaert	±1570–±1585
Michiel Coignet	1572/1573–1596
Melchior Blommaert	1587–1590
Hans van Weerden	1591–1605
Peter de Cock	1596–?
Gabriel van Bemel	1605–?

[a]In 1549–1550 he rented the wine excise for 93,992 gl. (s.n. 1864, p. 21)

The revenue that was divided between the wine gaugers in the period 1577–1593.

Table I

Year	Period	earnings	Source
1551		£84 4sch. 8gr. Brab.	R14 f354*r*
1577	16 January 1976–16 January 1977	87 gl. 9 st.	R18 f73*r*
1578	16 January 1977–2 January 1979	287 gl. 7 st.	R19 f74*r-v*
1579	2 January 1979–2 January 1980	148 gl. 10 st.	R20 f74*r*
1580	rekening niet op tijd binnen		R21 f74v–75*r*
1581	2 January 1980–1 January 1981	125 gl. 19 st.	R22 f80*v*–81*r*
	2 January 1981–1 January 1982	130 gl. 19 st.	
1582	1 January 1982–31 December 1982	129 gl. 19 st. 6 p.	R23 f71*r*
1583	1 January 1983–31 December 1983	103 gl. 14 st. 6 p.	R24 f71*v*–72*r*
1584	1 January 1984–21 December 1984	87 gl. 12 st.	R25 f79*r-v*
1585	no entries		R26
1586	1 January 1985–31 December 1985	21 gl. 9 st. 6 p.[a]	R27 f63*r-v*
	1 January 1986–31 December 1986	23 gl. 13 st.	
1588	1 January 1988–31 December 1988	72 gl. 16 st. 3 p.	R28 f67*r*
1589	1 January 1989–31 December 1989	87 gl. 9 st.	R29 f67*v*–68*r*
1590	1 January 1990–31 December 1990	84 gl. 0 st. 3 p.	R30 f62*r*
1591	1 January 1991–1 April 1991	26 gl.	R31 f58*v*
1592	earnings no longer divided by three		R32 f56*v*
	1 April 1991–2 April 1992	359 gl. 18st.	Pk1409 f43*r*
1593	2 April 1992–2 April 1993	265 gl.	Pk1409 f43*r*
1601	? voor Hans van Weerden	60 gl.	T527,
	voor Peter de Cock	90 gl.	kwitanties[b]

[a]"Waeraff de voirß wijnroeier (=Coignet) competeerende de twee derdendele" (two-thirds of which went to Michiel Coignet). Is this a reference to the fact that only Michiel Coignet served as a gauger?

[b]Both gaugers were paid a quarter of their earnings. Remarkably this cannot be traced in the archives of the city treasury

Archival Sources

In the course of researching this book, I consulted numerous documents in the Antwerp City Archives. Although not all are cited explicitly, they were invaluable to the descriptions provided on the foregoing pages. The following is a list of archival materials consulted. It is ordered by collections. The abbreviation between brackets is that by which they are cited in the book. During the writing of this monograph, the iconographic collection (Icon) was being renumbered, a process that had not been concluded by the time of publication. Anyone wishing to consult an item from this collection is therefore advised to contact the archivist for further information on document numberings.

General

L. BISSCHOPS, Genealogical notes – see catalogue LZ 26
F. DONNET, Genealogical notes – see catalogue LZ 26

Certificatieboeken – Certification Books (Cert.)

5, 9, 10, 11, 12, 15, 22, 23, 24, 25, 26, 30, 31, 33, 34, 36, 37, 38, 39, 40, 42, 47, 48, 50, 53.

Collectanea (Coll.)

12, 16, 38.

A. Meskens, *Practical Mathematics in a Commercial Metropolis: Mathematical Life in Late 16th Century Antwerp*, Archimedes 31, DOI 10.1007/978-94-007-5721-9,
© Springer Science+Business Media Dordrecht 2013

Copers en comparanten – Real Estate Transactions (CC – sample)

12–27.

Gilden en Ambachten – Guilds and Trades (GA)

4487, 4528, 4529, 4530, 4538, 4550, 4587, 4617, 4663, 4664, 4830(1), 4833.

Insolvente Boedelskamer – Estates (IB)

766
768 Memorieboeck Geraert Gramaye.
770–786 Nagelaten stukken Frans de Pape.
2899 (2040) Koopmansleerboek ± 1567. (Arithmetic book)
2964/MF096K M. Coignet, *Livre d'Arithmétique*, 1587.
(2174)

Iconografie – Iconography (Old Numbers Between Brackets)

12#5266 (3/3b) M. Coignet en M. Van Herle. Sint Michielsabdij.(saint Michael's Abbey)
12#9203 (14) S W. J. Delff. Portret F. Sambix.
15/12 J.B. Vrients. Beleg van Oostende (the Siege of Ostend).
15/13 J.B. Vrients. Beleg van Oostende (the Siege of Ostend).
15/14 J.B. Vrients. Beleg van Oostende (the Siege of Ostend).
15/16 Livius de Voghelare. Beleg van Oostende (the Siege of Ostend).
25/16 J. Ghysens. Erven tussen Roode Poort en Slijkpoort (Properties between the Red Gate and the Mud Gate).
26/1 Ontwerp versterking (design of fortifications).
26/2 Ontwerp versterking (design of fortifications).
26/2 Gezicht versterking (viewof fortifications).
26/3 B. Campi. Citadel van Antwerpen (Antwerp citadel).
26/3bis B. Campi. Citadel van Antwerpen (Antwerp citadel) .
26/4 Hans Vredeman de Vries. Incorporatie citadel in stadswallen (incorporation of the citadel into the city walls).
26/6
26/10 P. Frans. Plan citadel.
26/15d Citadel.
26/24bis M. Coignet (?). Vesten (moats).
36/34 Ontwerp vesten (design of moats).
30I/3 W.J. Blaeuw. Pascaarte van alle zeekusten van Europa 1625 (Portulan map of Europe's sea boards).
31/30b Tabula Castelli Sandflitam.

41/7 7bis 43A1ter Uitbreiding stad A. Andriessen en H. Vredeman de Vries (City extension plans)
64B/2c Oorlog Sluis (War Sluis)
64B/5c Forten 1584 (Fortifications 1584)
64B/5bis Forten 1585 (Fortifications 1585)
64B/5ter Forten 1585 (Fortifications 1585)
64B/6c Paskaart (Rutter)

Kerken en kloosters – Churches and Convents (K)

430.

Notarissen – Notaries (N)

466, 525, 535, 1329, 1478, 1480, 1481, 1483, 2071, 2074, 2077, 2875, 3377, 3378, 3408, 4675.

Parochieregisters – Parish Registers (PR)

1, 6, 64, 69.

Privilegekamer (Pk)

55 Lijste van de caerten, plans etc. berustende ter Librarije (list of maps in the library).
56 Plans 1er division, plans en portefeuille.
292, 477, 476, 481, 552, 554, 556, 557, 558, 628, 630, 634, 637, 638, 639 640, 641, 642, 643, 644, 645, 648, 649, 650, 651, 652, 655, 657, 658, 659, 660, 662, 663, 664, 666, 667, 668, 669, 672, 674, 675, 676, 677, 678, 679, 680, 681, 682, 683, 684, 686, 690, 691, 694, 695, 697, 699, 710, 712, 714, 715, 716, 717, 718, 719, 721, 905/1 t.e.m. 905/7, 917, 1510, 2566, 2766.
1409 Dossier wijnroeiers (winegauger's file).
2231 t.e.m. 2251 Fortificaties (fortifications).
2256, 2257, 2261, 2263, 2265, 2269, 2271, 2296, 2321.
2338, 2339, 2340 Groot Werck.
2633 Der Kooplieden Handboucxkin, J. Lambrecht, Gent, 1544 (Merchant's arithmetic book).

Processen – Trials

B1313, B2115, B2467, D6235, C2806, G7013, S100, S208, Suppl. 1739, Suppl. 5985, Suppl. 6327.

Rekeningen – Accounts (R)

14, 15, 16, 17, 18, 19, 20, 21, 22, 23, 24, 25, 26, 27, 28, 29, 30, 31, 32, 33, 34, 35, 36, 37, 38, 39, 63, 399, 1040, 1098, 1310bis, 1311, 1766, 1767, 1768, 1769, 1775, 1776, 1778, 1808, 1809, 1833, 1834, 1954, 2181, 2288, 2350, 2419, 2431, 2646.

Kohieren van 5de, 10de en 100ste penningen op huizen – Tax registers of the War Tax

2209, 2210, 2211, 2213, 2220, 2221, 2224, 2225, 2226, 2229, 2230, 2232, 2233, 2234, 2235, 2237, 2240, 2241, 2242, 2243, 2244, 2245, 2246, 2247, 2248, 2249, 2288, 2315, 2317, 2318, 2320, 2321, 2322.

Schepenregisters – Aldermen's Registers (SR)

156, 161, 248, 250, 253, 268, 272, 273, 276, 284, 288, 291, 292, 295, 296, 298, 299, 311, 312, 313, 314, 315, 317, 319, 320, 321, 322, 323, 326, 328, 329, 333, 336, 339, 341, 345, 348, 350, 360, 362, 363, 365, 366, 367, 370, 371, 372, 376, 379, 381, 384, 385, 386, 388, 389, 393, 396, 397, 399, 400, 402, 403, 406, 407, 409, 414, 418, 419, 420, 424, 430, 432, 434, 435, 437.

Tresorij – Treasury (T)

1/1, 6, 9, 47, 48, 50, 181–182, 191, 192, 198.
Wijnaccijns 108, 512, 515, 518, 521, 524, 527, 510, 513, 516, 519, 522, 525, 1221, 511, 514, 517, 520, 523, 526, 537.
Loterij 1431, 1432, 1433.

Vierschaar – High Court (V)

144, 146, 149, 152, 1100, 1104, 1107, 1170, 1176, 1252, 1389, 1391, 1393, 1397, 1399, 1401, 1403, 1404, 1405, 1406, 1407, 1409, 1440, 1493.

Weesmeesterskamer – Ophan's Chamber (W)

6, 18, 20, 25, 33, 35, 48, 76, 115, 118, 133, 145, 152, 174, 185, 201, 274, 281, 286, 293, 352, 356, 374, 440, 523, 885.

Instruments

Gillis Coignet the Elder

Astrolabe, 1556, Osservatorio Astronomico di Roma, Rome.
Astrolabe, 1557, 114 mm, Château d'Ecouen, Paris.
Astrolabe, 1557, 172 mm, mater only, private collection.
Astrolabe, 1560, 216 mm, Museum of the History of Science, Oxford.
Astrolabe, 1560, 184 mm, private collection.
Astrolabe, s.d. (ca. 1560), 183 mm, British Museum, London.
Equinoctial sundial, 1557, 91 mm, Koninklijke Musea voor Kunst en Geschiedenis, Brussels.
Compendium, 49 mm, private collection.
Sector, 1562 (!), 350 × 34 × 6 mm, Rijksmuseum, Amsterdam.[1]

Michiel Coignet

Astrolabe, 1572, 229 mm, Kunstgewerbemuseum, Berlin.
Astrolabe, 1598, 152 mm, Museo Naval, Madrid.
Astrolabe, 1601, 227 mm, Museum Boerhaave, Leiden.
Astrolabe, not signed, 1602, 230 mm, private collection.
Astrolabe, 1618, signed Michaelis Coigneti, Ferd. Arssenius and Iacobus de Costere, 406 mm, Museo Arqueológico, Madrid.
Astrolabe, not signed, not dated, 330 mm, Museum für Kunst und Gewerbe, Hamburg.
Astrolabe, not signed, not dated, 258 mm, Museum of the History of Science, Oxford.
Astrolabe, not signed, not dated, 270 mm, Castello Sforzesco, Milan.
Armillary sphere, 1591, diameter 358 mm, height 410 mm, Adler Planetarium, Chicago.
Armillary sphere, not signed, not dated, diameter ca. 500 mm, height ca. 1400 mm, Real Monasterio de San Lorenzo de El Escorial, Madrid.
Nocturlabe, not dated, 108 mm, height 165 mm, Museum of the History of Science, Oxford.
Nocturlabe-Astrolabe, not signed, not dated, 93 mm, height 167 mm, private collection.
Nocturlabe, not signed, not dated, 139 mm, height 235 mm, private collection.

[1]Although this date is engraved on the instrument, it is most likely erroneous. Otherwise the instrument would predate Michiel's first pantometers. Moreover, it bears a strong resemblance to the sectors described in his manuscripts and made by his workshop after 1610 (see Sect. 7.2).

Nocturlabe, not signed, not dated, 137 mm, height 233 mm, Royal Museum of Scotland, Edinburgh.
Table sundial, 1604, 80 mm height 115 mm, Geomagnetisches Institut, Potsdam.
Polyhedric sundial, height 280 × 165 × 165 mm, private collection.
Compendium, 1602, 70 mm, Sotheby's, London, 5 May 1989.
Holland circle, 1590, 178 mm, Drouot, Paris, 6 December 1985.
Holland circle, 1598, 195 mm, Christie's, London, 29 March 1990.
Holland circle, 1598, 167 mm, Drouot, Paris, 7 April 1987.
Holland circle, 1600, 162 mm × 176 mm, Château d'Ecouen, Paris.
Holland circle, 1602, 183 mm, Harvard University, Collection of Historical Scientific Instruments.
Holland circle, 1606, 177 mm, Adler Planetarium, Chicago.
Holland circle, 1606, Musée des Techniques, Parijs.
Holland circle, not signed, 1615, private collection.
Holland circle, not signed, not dated, 177 mm, Museum aan de Stroom, Antwerp.
Holland circle, not signed, not dated, Germanisches National-museum, Nuremberg.
Holland circle, not signed, not dated, 192 mm, with two alidades with pinhole views, private collection.
Holland circle, not signed, not dated, 194 mm, Koninklijke Musea voor Kunst en Geschiedenis, Brussels.
Compasses, not signed, not dated, 143 mm, private collection.
Pair of sectors, not signed, not dated, 180 × 34 × 4 mm (both), private collection.
Sector, not signed, not dated, 109 × 25 × 3 mm, private collection.
Protractor, not signed, not dated, half circle 142 mm, private collection.
Protractor, not signed, not dated, quarter circle, radius 77 mm, private collection.

Bibliography

Books and Manuscripts by Michiel Coignet

Books

Coignet, Michiel. *Almanach 1617*. s.n., 1616 (?). Stadsbibliotheek Kortrijk Almanakken 33.

Coignet, Michiel. *Almanach 1626*. s.n., 1625 (?). Stadsbibliotheek Kortrijk Almanakken 39.

Coignet, Michiel. *Cent questions ingenieuses et recreatives, pour delecter & aguiser l'entendement*. Antonius van Diest for Jan van Waesberghe, Antverpiae [Antwerpen], 1573. MPM A4348.

Coignet, Michiel. *Nieuwe Onderwijsinghe/op de principaelste Puncten der Zeeuaert*. Hendrick Hendricsen, Hantwerpen [Antwerpen], 1580. Appendix to: Pedro de Medina, *De zeevaert, oft: Conste van ter zee te varen*, EHC C1/16623.

Coignet, Michiel. *Instruction nouvelle des poincts plus excellents and necessaires, touchant l'art de naviger*. Henry Hendrix, Anvers [Antwerpen], 1581. MPM A 2727.

Coignet, Michiel. *Nieuwe Onderwijsinghe/op de principaelste Puncten der Zeeuaert*. C. Claesz, Amsterdam, 1589. University Library Gent R999.

Coignet, Michiel. *Nieuwe Onderwijsinghe/op de principaelste Puncten der Zeeuaert*. C. Claesz, Amsterdam, 1593. Minicipal Library Tournai.

Coignet, Michiel. *Nieuwe Onderwijsinghe/op de principaelste Puncten der Zeeuaert*. C. Claesz, Amsterdam, 1598. Nederlands Historisch Scheepvaartmuseum.

Coignet, Michiel. *L'uso del compasso di Fabrizio Mordente Salernitano & ... composto da Michele Coignetto Mathematico del Ser.mo Archiduca Alberto Austriaco*. s.n., Modena, 1608. Bibliotheca Estense Camp. 290 = Gamma U.1.18.

Coignet, Michiel. Tabula geographica indicans iter novum inter mediolanum et antverpiam. Abraham Verhoeven, 1621. Katholieke Universiteit Leuven, Tabularium, Eénbladsdrukken.

Coignet, Michiel. *M. Michel Connette, sur les propositions geometriques extraictes des six premiers liures des Elemens d'Euclide*. C. Hulpeau, Paris, 1626. with La geometrie reduite.

Coignet, Michiel, and P.G.S. (edit.). *La geometrie reduite en une facile et briefe pratique, par deux excellents instruments, dont l'un est le pantometre ou compas de proportion de Michel Connette, et l' autre est l'usage du compas a huict pointes invente par Fabrice Mordente compose en italien par M.C.... Traduits en francois par P.G.S. mathematicien*. C. Hulpeau, Paris, 1626. Museum of the History of Science, Oxford.

A. Meskens, *Practical Mathematics in a Commercial Metropolis: Mathematical Life in Late 16th Century Antwerp*, Archimedes 31, DOI 10.1007/978-94-007-5721-9,
© Springer Science+Business Media Dordrecht 2013

de Jode, Cornelis. *Speculum Orbis Terrarum* (with a mathematical introduction by Michiel Coignet). Arnoldum Coninx for Vidua et haeredes Gerardi de Judaeis, Antverpiae [Antwerpen], 1593. MPM R 45.3.

Henrion, Denis. *Deux cens questions ingénieuses et récréatives extraictes et tirées des oeuvres mathématiques de Valentin Menher, avec quelques annotations de Michel Coignet, le tout corrigé, recueilli et mis en cet ordre, par D. H. P. E. M.* s.n., Paris, 1620.

Henrion, Denis. *Collection ou recueil de divers traictez mathematiques.* F. Bourriquant, Paris, 1621. EHC G4787. Contains inter al. Devx cens qvestions ingenieuses et recreatives extraictes et tirées des oevvres mathematiqves de Vallentin Menher Allemand. Auec quelques annotations de Michel Coignet sur aucunes d'icelles questions. Le tout corrigé, recueilly & mis en cet ordre par DHPEM.

Mennher, Valentin, and Michiel Coignet (ed.). *Livre d'arithmétique, contenant plusieurs belles questions & demandes, propres & utiles a tous ceux qui hantent la trafique de marchandise.* Antonius van Diest for Jan van Waesberghe, Antverpiae [Antwerpen], 1573. MPM A4348.

Ortelius, Abraham, and Michiel Coignet (ed.). *Epitome theatri orbis terrarum.* Joannis Keerbergii, Antverpiae [Antwerpen], 1601. MPM R 61.15.

Ortelius, Abraham, and Michiel Coignet (ed.). *L'epitome du theatre de l'univers.* Joannis Keerbergii, Antwerpen, 1602. MPM 8–190.

Ortelius, Abraham, and Michiel Coignet (ed.). *Abraham Ortelius his Epitome of the Theatre of the worlde.* Ieames Shawe, London, 1603. British Library Maps C.2b10.

Ortelius, Abraham, and Michiel Coignet (ed.). *L'Epitome du theatre de l'univers d'Abraham Ortelius : nouvellement recogneu, augmentè, et restaurè de meseure geographique : par Michel Coignet mathemat[icien] d'Anvers.* Ioannisi Bapt. Vrintii, Antverpiae [Antwerpen], 1609. KBR IV 58.678 A (a).

Ortelius, Abraham, and Michiel Coignet (ed.). *Epitome theatri orbis terrarum.* officina Plantiniana, Antverpiae [Antwerpen], 1612a. MPM R 61.14.

Ortelius, Abraham, and Michiel Coignet (ed.). *Epitome theatri orbis terrarum.* officina Plantiniana, Antverpiae [Antwerpen], 1612b. R 61.14.

Ortelius, Abraham, and Michiel Coignet (introduction). *Theatrum Orbis Terrarum Abrahami Ortelii Antverp., geographii.* R. Bruneau for J.B. Vrients, 1609. University Library Groningen.

Ortelius, Abraham, and Michiel Coignet (introduction). *Theatrum Orbis Terrarum Abrahami Ortelii Antverp., geographii.* Officina Plantiniana, Antverpiae [Antwerpen], 1612. Koeman Ort 41.

Ortelius, Abraham, Michiel Coignet (ed.), and Filippo Pigafetta (trans.). *Theatro del mondo di Abrahamo Ortelio.* Giovanni Bapta. Vrinto, Anversa [Antwerpen], 1608. 1608.

Ortelius, Abraham, Michiel Coignet (ed.), and Filippo Pigafetta (trans.). *Theatro d'el orbe de la tierra ... el qual antes el estremo dia de su vida por la postrera vez ha emendado, y con nueuas tablas y commentarios augmentado y esclarescido.* la libreria Plantiniana, Anveres [Antwerpen], 1612a. MPM R 39.8.

Ortelius, Abraham, Michiel Coignet (ed.), and Filippo Pigafetta (trans.). *Theatro del mondo di Abrahamo Ortelio.* libreria Plantiniana, Anversa [Antwerpen], 1612b. 1608.

Raets, Willem, and Michiel Coignet. *Arithmetica, oft: Een niew cijfferboeck ... Met noch een tractaet vande wisselroede, met annotatien.* Hendrick Hendricsen, Antwerpen, 1580a. EHC G16053, MPM R50.27.

Raets, Willem, and Michiel Coignet. *Practycke om lichtelijck te leeren visieren alle vaten metter Wisselroede.* Hendrick Hendricsen, Antwerpen, 1580b. EHC G16053, MPM R50.27.

Raets, Willem, and Michiel Coignet. *Arithmetica, oft: Een niew cijfferboeck ... Met noch een tractaet vande wisselroede, met annotatien.* Hieronimus Verdussen, Hantwerpen [Antwerpen], 1597a. KBR III 93.159 A R.P.

Raets, Willem, and Michiel Coignet. *Practycke om lichtelijck te leeren visieren alle vaten metter wisselroede/Willem Raets ; Van nieus oversien... door Michiel Coignet.* Hieronimus Verdussen, Hantwerpen [Antwerpen], 1597b. KBR III 93.159 A 2 R.P.

Saminiatus, Federicus. *Tabulae astronomicae.* Martini Nutii, Antverpiae [Antwerpen], 1599. with an epigram by Michiel Coignet. MPM A2070, EHC G5004.

Manuscripts

Coignet, Michiel. Est de Michaelis Coigneti 1576, 1576–1577. part of the manuscript written 1603 and later; Bibliothèque Nationale Paris, Ms Néer 56.
Coignet, Michiel. Della forma et Parti del Compasso di Fabritio Mordente salernitano. Con gli usi di esso, raccolti da Michele Coignet Mathematico del Serenissimo Archiduca Alberto. Per quali si risolvono molte propositioni, cavate dalla primi sei libri d'Euclide, 1608. Bibliotheca Estense Modena, MS Gamma G.4.34 (Campori, 548).
Coignet, Michiel. Usus duodecim diuisionum geometricarum, 1610–1612. KBR MS II769.
Coignet, Michiel. Usus trium praecipuorum Mathematicorum instrumentorum, 1612. Bibliotheca Nacional Madrid, MS 9213.
Coignet, Michiel. Descrittione amplissa dell' uso della dioptra Geometrica, 1613. City Library Modena, MS A1844.
Coignet, Michiel. Beschreibung des Passers Fabricii Mordenti von Salermo..., aus dem Italianischen in Deutsch vertirt. Prag. Geschrieben zu Berlin, Anno 1616., 1616. Bibliothèque Nationale Paris, Allemand 159 (cote) Supplément français 3937[44].[1]
Coignet, Michiel. Plan of the Antwerp canals, 1618a. SAA Icon26/24bis.
Coignet, Michiel. El uso de las doze diuisiones geometricas, 1618b. EHC B264708. Contains five treatises: De la composicion y uso de las dos Reglas pantometras; El uso del compas Proporcional; El uso de la Dioptra geometrica, y de la esquadra para d'Artilleria; Sigue la platica para medir distancias por muchas maneras particulares; Sigue el uso del Quadrado geometrico Hamado tambien Escala geometrica.
Coignet, Michiel. Circini Fabricii Mordenti Salermonensis descriptio, ex Germanico idimate in Latinum translata, anno 1624., 1624. with Coignet (1616).
Coignet, Michiel. De Regulae Pantometrae fabrica & usu libri septem, after 1604. Bodleian Library Oxford, MS Canon Misc 243.
Coignet, Michiel. Livre d'Arithmetique escrit l'an 1587 et apres reueu et emende par l'auteur M. Michiel Cognet Arithmeticien et Geometre tressingulier de la ville d'Anuers, ca. 1587. SAA IB2964.
Coignet, Michiel. De Regula Pantometre fabrica & usu libri septem, ca. 1604. Bibliothèque Nationale Paris, J253.
Coignet, Michiel. Michaelis Coigneti Antverpiani Duodecim divisionum regulae pantometrae, per quas ope vulgari circini omnia mathematicorum problemata facili negotio resolvuntur, ca. 1610a. Národní Knihovna (National Library) Prague, XII.A.12.
Coignet, Michiel. Usus regulae pantometrae, ca. 1610b. Florence, MS Riccardiana 956.
Coignet, Michiel. El Usso del Compas Proporcional compuesto par Miguel Connietto, ca. 1610c. Bibliotheca Nazionale Napels, MS.I.D.I.
Coignet, Michiel. Usus duodecim divisionem regulae pantometrae, ca. 1610d. Bibliotheca Vaticana, Ms Reginiensis Latinus 1307.
Coignet, Michiel. Usus duodecim divisionem regulae pantometrae, ca. 1610e. Florence, Bibliotheca Riccardiana MS 859.
Coignet, Michiel. Uso del compasso di Fabricio Mordenti di salerno Mathematico del. ser. mo Principe Allesandro Farnese Ducca di Parma composto per Michalelo Coignetto, s.d.a. Bibliotheca Nacional Madrid, MS 19.709/32.
Coignet, Michiel. Neunspitziger Passer. Fabricii Mordenti von Salermo Mathematici des Herzogen Alexandrii Fernesi Herzogen zu Parma., s.d.b. University Library Wrocław Ms R461.
Coignet, Michiel, and Mattheus van Herle. Fortifications near St-Michael's Abbey, s.d. SAA Icon 3/3b.
Coignet, M. (ascribed). Fabrica y uso del pantometro, s.d. AFL 28.3.
s.n. Manual for the use of Coignet type sectors, with notes by Joannes della Faille, ca. 1620. private collection.

[1]See Huet (1894–1895, p. 69).

Unpublished Sources

Anselmo, A.A. Album amicorum, 1594–1602. KBDH 71 J 57.
de Burbure, L. Archiefnota's. SAA Pk2932–2934, s.d.
de Groote, H.L.V. Nagelaten papieren (Fonds H.L.V. De Groote). SAA Pk3470–3496, s.d.
Dilis, E. Nota's goud en zilversmeden. SAA Pk3211, s.d.
Goris, J.A. Nota's Schetz. SAA Pk3449, s.d.
Heyns, P. Rekenboek 1576–1585, 1576. MPM M394.
Heyns, P. Rekenboek 1580 1584, 1580. MPM M240.
Heyns, P. Memorieboek, 1583. MPM M73.
le Poivre, P. Collections of plans, s.d. KB Hs 19611.
Nouts, H. Rekenschrift, s.d. Nederlands Economisch Historisch Archief NO351.
Proost, J. Inventaire ou table alphabétique des noms de personnes contenus dans les registres aux gages et pensions des chambres des comptes. Algemeen Rijksarchief, 1890.
Roggeman, G. Notaris Adolf van de Venne dl. 1, 1579–1605. Stadsarchief Mechelen, s.d.a.
Roggeman, G. Notaris Adolf van de Venne dl. 2, 1606–1615. Stadsarchief Mechelen, s.d.b.
Rooses, M. Geschiedenis der Antwerpsche Schilderschool. SAA Pk3409, 1879.
Sacrobosco. De Sphaera, s.d. MPM M326.
s.n. Stedelijke archieven van Antwerpen, Weesmeesterskamer, Tafel der Registers van 1576–1658. Stadsarchief Antwerpen, 1871.
s.n. Antwerpse poortersboeken 1533–1608. Stadsarchief Antwerpen, 1977.
s.n. Koopmansleerboek (Merchant's manual), s.d.a. MPM M78.
s.n. Koopmansleerboek (Merchant's manual), s.d.b. MPM M318.
s.n. Wisselkoersen (Exchange rates), s.d.c. MPM M91.
s.n. Astrological treatise, s.d.d. MPM M51.
s.n. Gebruiksaanwijzing sferisch astrolabium.(Manual for the use of a spherical astrolabe), s.d.e. MPM M385.
van den Branden, F.J. Nota's alfabetisch. SAA Pk3359, s.d.a.
van den Branden, F.J. Nota's betreffende Van Woelput, Jan Ympen... SAA Pk3358, s.d.b.
van den Branden, F.J. Nota's i.v.m. schilders. SAA Pk3571–3580, s.d.c.
van Halle, P. Arithmetice, s.d. KBR Hs 3552.
van Hemeldonck, G. Kunstenaars 1500–1600, Ontledingen van archiefteksten. SAA Pk3524, 1987.
van Hemeldonck, G. *Repertorium der Belgische goud- en zilversmeden.* digital archive of Antwerp City Archives, 1998.
van Roey, J. Fiches in verband met de maandelijkse quotisatie, Leeszaal. Stadsarchief Antwerpen, s.d.
Vivianus, J. Album Amicorum, 1571–. KBDH 74F19.
Wentsel, M. Instrucsye op het Italiaens bouckhouden, ca. 1588. Technische Universiteit Delft.
Zael, D. [?]. [Carte marine et instructions nautiques pour la navigation de Helgoland jusqu'au sud du Portugal], ca. 1550? EHC B 29166 [C2–544 C].

Published Sources

Aguilonius, F. 1613. *Opticorum libri sex.* ex officina Plantiniana, apud viduam et filios Jo. Moreti. Antverpiæ [Antwerpen]: Plantin.
Alberi, E. 1853. *Le Opere di Galileo Galilei.* Firenze: Societa Editrice Fiorentina.
Andreas, V. 1643. *Bibliotheca Belgica.* Lovanii [Leuven]: Jacobis Zegeris.
Arias Montanus, B. 1573. *Divinarum nuptiarum conventa et acta.* Antverpiae [Antwerpen]: Gillis Coppens van Diest.

Baldini, U. (ed.). 1995. *Christoph Clavius e attività scientifica dei gesuiti nell'eta di Galileo*. Roma: Bulzoni Editore.

Barlow, W. 1597. *The navigator's supply*. London: G. Bishop.

Barrefelt, H. ca.1592. Bibelsche figuren [Franciscus I Raphelengius], [Leiden].

Blaeu, W.J. 1631. *Zeespiegel, inhoudende een korte onderwijsinghe inde konst der zeevaert, en beschryvinghe der seen en kusten van de oostersche, noordsche, en westersche schipvaert*. Amsterdam: Willem Jansz Blaeuw.

Blundiville, M. 1594. *His exercises, containing sixe treatises*. London: Iohan Windet.

Borough, W. 1581. *Of the variation of the compas*. London: s.n.

Bruslé de Montpleinchamp, J. 1693. *L'histoire de l'archiduc Albert*. Köln: Erven C. Egmond.

Chuquet, N., and H. l'Huillier (ed.). 1979. *La Géométrie, Première géométrie algébrique en langue française (1484)*. Paris: VRIN.

Clavius, C.1608. *Algebra*. Rome: Bartollomeum Zannetum.

Collado, L. 1606. *Practica manuale di Arteglieria*. Milano: G. Bordoni.

Copernicus, N., and S. Hawking (ed.). 2002. *On the revolutions of the heavenly spheres*. Philadelphia: Running Press.

de Medina, P. 1580. *De zeevaert, oft: Conste van ter zee te varen*. Hantwerpen [Antwerpen]: Hendrick Hendricsen.

Dodoens, R. 1548. *Cosmographica in astronomiam et geographiam isagoge*. Antverpiae [Antwerpen]: ex officina Joannis Loëi [Loe, van der, Jan].

Eytzinger, M. 1583. *De Leone Belgico, ejusque topographica atque historica descriptione liber*. Coloniae Ubiorum [Köln]: Gerardus Campensis.

Eytzinger, M. 1585. *De Leone Belgico, ejusque topographica atque historica descriptione liber*. Coloniae Ubiorum [Köln]: Gerardus Campensis.

Foppens, J.F. 1739. *Bibliotheca Belgica, sive: Virorum in Belgio vita*. Bruxellis [Brussel]: P. Foppens.

Frisius, G. 1533. *Libellus de locorum describendorum ratione, & de eorum distantiis inveniendis nunquam ante hac viene*. Antverpiae [Antwerpen]: Birckman.

Frisius, G. 1540. *Arithmeticae practicae methodus facilis*. Antverpiae [Antwerpen]: vaeneunt Gregorio Bontio [Gregorius de Bonte]; typis Aegidii Coppenii Diesthen. [Gillis I Coppens van Diest].

Frisius, G. 1553. *De principiis astronomiae et cosmographiae, deque usu globi*. Antverpiae [Antwerpen]: ædibus Joannis Steelsii [Joannes Steelsius].

Frisius, G. 1556. *De astrolabo catholico liber quo latissime patentis instrumenti multiplex usus explicatur, & quicquid uspiam rerum mathematicarum tradi possit continetur*. Antverpiae [Antwerpen]: aedib. Joan. Steelsii [Joannes Steelsius], MPM A 2429.

Frisius, G., and P. Apianus. 1533. Cosmographicus liber ... item ... libellus de locorum describendorum ratione, & de eorum distantiis inveniendis. Antverpiae [Antwerpen]: Vaeneunt ... per Arnoldum Birckman [Birckman, Arnold I]; Joan. Grapheus typis cudebat [Grapheus, Joannes].

Frisius, G., and P. Apianus. 1574a. *Cosmographia*. Antverpiae [Antwerpen]: apud Greg. Bontiu[m] [Gregorius de Bonte].

Frisius, G., and P. Apianus, 1574b. *Cosmographia*. Antverpiae [Antwerpen]: apud Christophorum Plantinum [Christoffel Plantijn].

Galileï, G. 1968. *Le Opere di Galileo Galilei*, 20 vols. Florence: G. Barbera editore.

Gallucci, G.P. 1597. *Della fabrica et uso di diversi stromenti di astronomia et cosmografia, ove si vede la somma della teorica et pratica di queste nobilissime scienze*. Venetia [Venezia]: appresso Ruberto Meietti.

Antonius Gillis, M. 1564. *Hantboecxken leerende na der stoischer Philosophïe wyse hoe elc in sinen roep gherustelyck leven sal*. Antwerpen: Gillis Coppens van Diest voor Jan van Waesberghe.

Guicciardini, L. 1567. *Description de tout le Païs-Bas autrement dict la Germanie Inferieure, ou Basse-Allemaigne*. Anvers [Antwerpen]: W. Sylvius.

Guicciardini, L. 1582. *La description de tous les Pais-Bas, autrement appellés la Germanie inférieure, ou Basse Allemagne*. Anvers [Antwerpen]: De l'imprimerie de Christophle Plantin.
Guicciardini, L. 1979. *Antwerpen, Mechelen, Lier e.a. (vertaald door Kiliaan)*. Soethoudt, Deurne.
Heyns, P. 1560. *Jaerdicht van tnieu stadthuys van Antwerpen*. Antwerpen: s.n.
Heyns, P. 1561. *Tot profijte van die willen leeren*. Antwerpen: s.n.
Heyns, P. 1568. *ABC, oft Exemplen om de kinderen bequamelick te leeren schryven*. Antwerpen: Christoffel Plantyn.
Heyns, P. 1581. *Feest-dicht voor Gilis Hooftman*. Antwerpen: [C. Plantin].
Heyns, P., and E. Ruijsendaal (eds.). 2006. *Cort onderwijs van de acht deelen der Fransoischer talen (1571 en 1605), volume 1 of Geschiedenis van het talenonderwijs in de Lage Landen*. Amsterdam: Stichting Neerlandistiek VU.
Hondius, H. 1624. *Korte beschrijvinge, ende af-beeldinge van de generale regelen der fortificatie, de artillerie, munition, ende vivres, van de officieren der selver ende hare commissien, van de leger-aerde-wallen, de approchen met het tegenweer, ende van vyverwercken*. ex officina Henr. Hondii [Hendrik Hondius], Hagae-Comit. ['s-Gravenhage].
Huet, G. 1894–1895. *Catalogue des manuscrits allemands de la Bibliothèque nationale*. Paris: Librairie Emile Bouillon.
Kepler, J. 1938–. *Gesammelte Werke*, 20 vols. München: C.H. Beck'sche Verlagsbuchhandlung.
Marolois, S. 1614–1615. *Opera mathematica, ou: OEuvres mathematiques traictans de geometrie, perspective, architecture, et fortification*. H. Hondius, Hagae-Comitis ['s-Gravenhage].
Mennher, V. 1556. *Arithmétique seconde*. Anvers [Antwerpen]: par Jan Loë [Jan van der Loe].
Mennher, V. 1561. Livre d'arithmetique, contenant plusieurs belles questions et demandes, propres et utiles à tous marchans. Anvers [Antwerpen]:Gilles Copenius de Diest.
Mennher, V. 1563. *L'arithmetique ... contenante plusieurs belles questions iournellement occurrentes en trafficque de marchandise*. Anvers [Antwerpen]: Gilles Copenius de Diest.
Mennher, V. 1565. *Practique pour brievement apprendre a ciffrer, & tenir livre de comptes*. Anvers [Antwerpen]: Aegidius Diest [Coppens van Diest, Gillis I].
Mercator, G., and J. Hondius (ill). ca.1607. *Atlas minor*. Amsterodami [Amsterdam]: excusum in aedibus Judoci Hondii [Jodocus I Hondius]; apud Cornelium Nicolai [Cornelis Claeszoon]; apud Joannem Janssonium [Johannes I JanssoniusI].
Mersenne, M. 1636–1637. *Harmonie universelle, contenant la théorie et la pratique de la musique*, vol. 2. Paris: S. Cramoisy.
Metius, A. 1623. *Praxis nova geometrica per usum Circini et regulae proportionalis*. Franekerae [Franeker]: U. Balck.
Metius, A. 1626. *Maet-constigh liniael ofte Proportionalen ry ende platten passer*. Franeker: U. Balck.
Mordente, F. 1567. *Modo di trouare con l'astrolabio, ò quadrante, ò altro instromento, oltre gradi, inticri, i minuti et secondi, et ognaltra particella*. Venezia: Bodleian Library.
Mordente, F., and G. Mordente. 1591. *La quadratura del cerchio: la scienza de'residui/il compasso et riga di Fabritio et di Gasparo Mordente*. Anversa [Antwerpen]: Philippus Galle.
Ortelius, A. 1612. *Theatrum Orbis Terrarum*. Antverpiae [Antwerpen]: Officina Plantiniana.
Ortelius, A. 1968. Album Amicorum. *De Gulden Passer* 46: 12–99.
Ortelius, A., and P. Heyns. 1579. *Le miroir du monde ... premierement en rithme brabançonne, par M.P. Heyns; et maintenant tourné en prose françoise*. Anvers [Antwerpen]: de l'imprimerie de Christophle Plantin pour Philippe Galle.
Ortelius, A., and P. Heyns. 1583. *Spieghel der werelt, ghestelt in ryme door M. Peeter Heyns ... Met een cort verhael van t'ghene ... van ... Anjou hier ... gheschiedt is*. Antwerpen: Christoffel Plantijn voor Philips Galle.
Ortelius, A., and R.A. Skelton (eds.). 1968. *The theatre of the world*. Amsterdam: Theatrum orbis terrarum.
Pisani, O. 1613. *Astrologia seu motus, et loca siderum*. Antverpiae [Antwerpen]: ex officina Roberti Bruneau.

Portantius, J. 1577. *Beschrijvinghe der nieuwer cometen, met aenwijsinge ende vermaninge wat zy bedieden ende voortbrenghen sal*. Antwerpen: François van Ravelenghien [Raphelengius, Franciscus I] and Christoffel Plantijn.

Puteanus, E. 1619. *De cometa anni M.DC.XVIII. novo mundi spectaculo, libri dvo. Paradoxologia.* Lovanii [Leuven]: apud Bernardinum Masium [Masius, Bernardinus I].

Riccioli, G.B. 1651. *Almagestum novum astronomiam veterem novamque complectens*. Bononiae [Bologna]: Ex typotraphia haeredis Victorij Benatij.

Schedel, H. 1493. *Liber cronicarum cum figuris et ymaginibus ab inicio mundi*. Nuremberge [Nürnberg]: Anthonius Koberger.

Scribani, C. 1610. *Origines Antverpiensium*. Antverpiae [Antwerpen]: Plantin.

Stevin, S. 1582. *Tafelen van interest, midtsgaders de constructie der selver*. Antwerpen: Christoffel Plantijn.

Stevin, S. 1617. *Nieuwe maniere van sterctebou, door spilsluysen*. Rotterdam: Jan van Waesberghe [Waesberge, van, Jan II].

Sweertius, F. 1628. *Athenea Belgica*. Antverpiae [Antwerpen]: G. a Tungris.

Tacquet, A. 1707. *Opera mathematica*, 2nd ed. Antverpiæ[Antwerpen]: Henricus and Cornelius Verdussen.

Ufano, D. 1614. *Artillerie*. Frankfurt: E. Emmel.

van Ceulen, L. 1584. *Solutie ende Werckinghe op twee geometrische vraghen by Willem Goudaen Inde Jaren 1580 ende 83, binnen Haerlem aenden Kerckdeure ghestelt. Mitsgaders Propositie van twee andere Geometrische vraghen tsamen door Ludolph van Colen gheboren in Hildesheim*. Amsterdam: Cornelis Claesz.

van Mander, C., A.F. Mirande, and G.S. Overdiep (eds.). 1943. *Het Schilder-Boek: het leven der doorluchtige Nederlandsche en Hoogduitsche schilders, naar de tweede druk 1618*. Amsterdam: Wereldbibliotheek.

van Mander, K. 1969. *Het schilder-boeck (facsimile van de eerste uitgave, Haarlem 1604)*. Utrecht: Davaco Publishers.

Van Roomen (Romanus), A. 1593. *Ideae Mathematica pars prima*. Antwerpen: J. van Keerberghen.

vanden Dijcke, M. 1591/92. *Soo behoort men te doen. Den Reghel van gevvin ende verlies vander vvaere*. Antwerpen: s.n.

vanden Dijcke, M. 1600. *Instrvctie om de Vvijnroede ende Peghelstock te maken/daer mede men oock sonder Peghelstock terstont meten en weten can de grootte van alle Vaten ende Tonnen. [...] Nv int licht gebracht door Martin vanden Dijcke, Meester van Chijfferen, Rekenen, ende Boeckhoude, binnen de vermaerder Coopstadt van Antwerpen*. Antwerpen: A. de Ballo. MPM R37.4(22).

vanden Hoecke, G. 1545. *In arithmetica: een sonderlinge excellent boeck, leerende veel schoone ende perfecte regulen der selver conste*. Thantwerpen [Antwerpen]: Symon Cock.

Verhoeven, A. 1621. *Oprechte afbeeldinge van zijn excellencie den marquis Ambrosius Spinola*. Hantwerpen [Antwerpen]: A. Verhoeven.

Verniers, J. 1592. *Almanach ende prognosticatie*. Thantwerpen [Antwerpen]: J. Trognesius.

Wagenaer, L.J. 1584. *Teerste deel vande spieghel der zeevaerdt*. Leyden [Leiden]: Christoffel Plantijn.

Cited References

Alden, D. 1996. *The making of an enterprise. The society of Jesus in Portugal, its Empire and beyond 1540–1750*. Stanford: Stanford University Press.

Anderson, R.G.W. 1972. *The mariner's astrolabe*. Edinburgh: Royal Scottish Museum.

Anon. 1863. *Antwerpsche parochiekerken, volume 2 of Verzameling der graf- en gedenkschriften van de provincie Antwerpen*. Antwerpen: J.E. Buschman.

Anon. 1864. De rekening van de Stad Antwerpen 1549–1550. *Antwerpsch Archievenblad* 1: 17–104.

Anon. 1964. *Het beleg van Hulst*. Utrecht: Stichting Menno van Coehorrn.

Anon. 1985. *The Planispheric astrolabe*. Greenwich: National Maritime Museum.

Arrighi, G. 1960. Benedetto Saminiati e il manoscritto 2723 della Bibliotheca Governativa di Lucca. *Physis* 2: 325–329.

Asaert, G. 2005. *Honderd huizen aan de Grote Markt van Antwerpen: vijf eeuwen bewoningsgeschiedenis*. Zwolle: Waanders.

Baetens, R. 1976. *De nazomer van Antwerpens welvaart; de diaspora en het handelshuis De Groote tijdens de eerste helft der zeventiende eeuw, volume 45 of Historische Uitgaven Pro Civitate*. Brussel: Gemeentekrediet van België.

Baldini, U., and P.D. Napolitani (eds.). 1992. *Christoph Clavius Corrispondenza. Edizione critica. V (1602–1605)*, vol. 5. Pisa: Dipartimento di Matematica.

Barnard, F.P. 1981. *The casting counter and the counting board*. Castle Cary: Fox.

Bennett, J.A. 1996. *The measurers. A Flemish image of mathematics in the sixteenth century*. Oxford: Museum of the History of Science.

Bennett, J.A., and S. Johnston. 1996. *The geometry of war 1500–1750*. Oxford: Museum of the History of Science.

Bertels, I., and P. Martens. 2006. Defunct defenses: Antwerp's sixteenth century defences. *Future Anterior* III(1): 44–53.

Beterams, G. 1939. Een halve eeuw humanistisch onderwijs te Mechelen. De Grootschool van 1585 tot 1635. Ph.D. thesis, Katholieke Universiteit Leuven.

Bloemendal, J. 2003. *Spiegel van het dagelijks leven? Latijnse school en Toneel in de noordelijke Nederlanden*. Hilversum: Verloren.

Bockstaele, P. 1970. De visierroede. Bibliografische bijdrage tot de geschiedenis van een oud meetinstrument. In *Handelingen van het XLIde Congres van de Federatie van Kringen voor Oudheidkunde en Geschiedenis in België*, ed. Anon, 526–537. Mechelen.

Bockstaele, P. 1976. The correspondence of Adriaan van Roomen. *Lias* 3: 85–129 and 249–299.

Bockstaele, P. 1985. Gielis van den Hoecke en zijn 'sonderlinghe boeck in dye edel conste Arithmetica'. *Academia Analecta, Klasse der Wetenschappen* 47: 1–29.

Bockstaele, P. 1989. Astrologie te Leuven in de zeventiende eeuw. *De Zeventiende Eeuw* 5: 172–181.

Bond, J.D. 1921. The development of trigonometric methods down to the close of the 15th century. *ISIS* 4: 295–323.

Bornhöft, E. 1985. Urkundliche Belege zum Lübecker Weinhandel. In *Lübecker Weinhandel: Kultur- und Wirtschaftsgeschichtliche Studien, Kultur- und Wirtschaftsgeschichtliche Studien*, 29–40. Lübeck: Amt für Kultur.

Bosmans, H. 1900. Le Traité des Sinus. *Annales de la Societé Scientifique de Bruxelles* 25: 91–121.

Bosmans, H. 1903. Documents inédits sur Grégoire de Saint-Vincent. *Annales de las Societé Scientifique* 27: 43–44.

Bosmans, H. 1908. La 'Practique om te leeren cypheren' de Nicolas Petri de Deventer. *Annales de la Societé Scientifique de Bruxelles* 32: 272–301.

Bosmans, H. 1909. Sur les papiers de l'astronome Wendelin aux archives générales du royaume à Bruxelles. *Annales de la Societé Scientifique de Bruxelles* 33: 74–92.

Bosmans, H. 1928. Théodore Moretus de la Compagnie de Jésus, Mathématicien (1602–1667). D'après sa correspondance et ses manuscrits. *De Gulden Passer* 6: 57–163.

Bostoen, K. (ed.). 1999. *Het album J. Rotarii: tekstuitgave van het werk van Johan Radermacher de Oude (1538–1617)*. Hilversum: Verloren.

Boumans, R. 1952a. De getalsterkte van katholieken en protestanten te Antwerpen in 1585. *Belgisch Tijdschrift voor Philologie en Geschiedenis* 6: 109–127.

Boumans, R. 1952b. Was Abraham Ortelius katholiek of protestant. *Handelingen der Zuidnederlandse Maatschappij voor Taal- en Letterkunde en Geschiedenis* 6: 109–127.

Boumans, R. 1965. *Het Antwerpse stadsbestuur voor en tijdens de Franse overheersing*. Brugge: De Tempel.
Bourland, C.R. 1951. *The Guild of St-Ambrose, or Schoolmaster's Guild of Antwerp 1529–1579*. Smith College Studies in History 36. Northampton: Smith College.
Briels, J.G.C.A. 1971. Zuidnederlandse goud- en zilversmeden in Noord-Nederland omstreeks 1576–1625, Bijdrage tot de kennis van de Zuidnederlandse immigratie. *Bijdragen tot de Geschiedenis* 54: 87–141.
Briels, J.G.C.A. 1972a. Zuidnederlandse onderwijskrachten in Noord-Nederland 1570–1630. *Archief voor de Geschiedenis van de Katholieke Kerk in Nederland* 14: 89–169.
Briels, J.G.C.A. 1972b. Biografische aantekeningen betreffende Zuidnederlandse onderwijskrachten in Noord-Nederland 1570–1630. *Archief voor de Geschiedenis van de Katholieke Kerk in Nederland* 14: 277–298.
Briels, J.G.C.A. 1973. Biografische aantekeningen betreffende Zuidnederlandse onderwijskrachten in Noord-Nederland 1570–1630. *Archief voor de Geschiedenis van de Katholieke Kerk in Nederland* 15: 103–149 and 263–297.
Briels, J.G.C.A. 1980. Zuidnederlandse immigranten en onderwijs in de Republiek omstreeks 1572–1630. *Spiegel Historiael* 15: 649–657.
Brokken, A. 1994. *Sterren in beelden*. St-Niklaas: Stadsbestuur en vzw Mercator.
Brouwers, L. 1972. *Brieven van Carolus Scribani*. Antwerpen: Vereniging Antwerpsche Bibliophielen.
Brown, J. 1979. Guild organisation and the instrument making trade 1550–1830: The grocers. *Annals of Science* 36: 1–34.
Brulez, W. 1959. *De firma della Faille en de internationale handel van Vlaamse firma's in de zestiende eeuw, volume 35 of Klasse der Letteren*. Brussel: Verhandelingen van de K.V.A.W.L.S.K. van België.
Brulez, W. 1960. De diaspora der Antwerpse kooplui op het einde van de 16de eeuw. *Bijdragen tot de Geschiedenisis der Nederlanden* 15: 279–306.
Bryden, D.J. 1992. Evidence from advertising for mathematical instrument making in London 1556–1714. *Annals of Science* 49(4): 301–336.
Burger, C.P. 1908. Oude Hollandse Zeevaartuitgaven, de oudste leeskaarten. *Tijdschrift voor Boek- en Bibliotheekwezen* 6: 245–261.
Burger, D. 1964. *Galileo Galileï*. Zeist; De Haan.
Cajori, F. 1993. *A history of mathematical notations*. New York: Dover.
Camerota, F. 2000. *Il Compasso di Fabrizio Mordente, volume 37 of Biblioteca di Nuncius*. Firenze: Leo S. Olschki.
Carli, A., and A. Favaro. 1896. *Bibliografia Galileiana (1568–1895)*. Roma: Bencini.
Caullet, M.G. 1904. Une collection d'almanachs placards (1560–1786). *Bulletijn van den geschied- en Oudheidkundigen Kring van Kortrijk* 2: 39–95.
Chaineux, M.C. 1981. *Culture de la vigne et commerce du vin dans la region de Liège au Moyen Age*. Liège-Leuven: CBHR-BCLG.
Chapman, A. 1983. A study of the accuracy of scale graduations on a group of European astrolabes. *Annals of Science* 40: 473–488.
Chomarat, M., and J.P. Laroche. 1989. *Bibliographie Nostradamus XVIe, XVIIe, XVIIIe siècles*. Baden-Baden Bouxwiller: V. KOerner.
Clair, C. 1955. A secret of the laurel-tree unveiled: another "rekenboeck" comes to light. *De gulden passer/Vereeniging der Antwerpsche Bibliophielen* 33: 36–41.
Couvreur, W. 1985. Galle en Hoefnagels stadsplattegronden en de Antwerpse verdedigingswerken van september 1577 tot februari 1581. In *Liber Amicorum L. Voet*, ed. F. de Nave, 519–545. Antwerpen: Vereniging Antwerpsche Bibliophielen.
Craeybeckx, J. 1958. *Un grand commerce d'importation: les vins de France aux anciens Pays-Bas (XIIIe-XVIe siècle)*. Paris: S.E.V.P.E.N.
Crone, E. 1953. Pedro de Medina, zijn leerboek der stuurmanskunst en zijn invloed op de cartografie in de lage landen. *Tijdschrift van het Koninklijk Nederlands Aardrijkskundig Genootschap* 70: 467–471.

Crone, E. 1966. De vondst op Nova Zembla. *Bulletin van het Rijksmuseum* 14: 71–85.

Dacos, N. 1995. Gillis Cognget. In *Fiamminghi a Roma*, 157–161. Brussel: Snoeck-Ducaju.

Davids, C.A. 1986. *Zeewezen en wetenschap: de wetenschap en de ontwikkeling van de navigatietechniek in Nederland tussen 1585 en 1815*. Amsterdam: Bataafsche Leeuw.

De Bougnge, L. 1952. *Bibliographie van kan. dr. F. Prims*. Antwerpen: De Sikkel.

de Groof, B. 1995. *Alexander Farnese and the origins of modern Belgium*. Leuven: University of Leuven.

De Groote, H.L.V. 1960. 16de eeuwse Antwerpse boekhoudkundigen en cijfermeesters. *Scientiarum Historia* 2: 161–172.

De Groote, H.L.V. 1961. 16de eeuwse Antwerpse boekhoudkundigen en cijfermeesters. *Scientiarum Historia* 3: 142–156.

De Groote, H.L.V. 1967. De 16de eeuwse Antwerpse schoolmeesters. *Bijdragen tot de Geschiedenis* 19: 179–318.

De Groote, H.L.V. 1968a. Een handschrift over rekenkunde van Michiel Coignet. hoe berekende de Brusselse rekenkamer een annuïteit in 1538?. *Scientiarum Historia* 10: 199–207.

De Groote, H.L.V. 1968b. De 16de eeuwse Antwerpse schoolmeesters. *Bijdragen tot de Geschiedenis* 20: 5–52.

De Groote, H.L.V. 1968c. Michiel Coignet. In *Nationaal Biografisch Woordenboek 3*, 184–191. Brussel: Koninklijke Vlaamse Academie van België.

De Groote, H.L.V. 1970a. Nog een zestiende-eeuws rekenboekje teruggevonden: de 'Cent Questions ingenieuses' van Michiel Coignet. *Scientiarum Historia* 12: 159–163.

De Groote, H.L.V. 1970b. Jan de Raeymaker. In *Nationaal Biografisch Woordenboek 4*, 723–726. Brussel: Koninklijke Vlaamse Academie van België.

De Groote, H.L.V. 1971. De 16de-eeuwse Nederlandse drukken over boekhouden en handelsrekenen hoofdzakelijk in betrekking met Antwerpen. *De Gulden Passer* 49: 6–28.

de Jong, M.A.G. 2005. *Staat van oorlog (1585–1621)*. Hilversum: Verloren.

de la Solla Price, D.J. 1955. An international checklist of astrolabes. *Archives Internationalesd'Histoire des Sciences* 8: 243–263 and 363–383.

de Lettenhove, K. 1909. *Albert et Isabelle (1595–1632)*. Brussel: Van Oest.

de Maeyer, M. 1955. *Albrecht en Isabella en de schilderkunst: bijdrage tot de geschiedenis van de 17e-eeuwse schilderkunst in de Zuidelijke Nederlanden, volume 8 of Verhandelingen van de Koninklijke Vlaamse Academie voor Wetenschappen, Letteren en Schone Kunsten van België: Klasse der Schone Kunsten*. Brussel: Koninklijke Vlaamse Academie voor Wetenschappen, Letteren en Schone Kunsten van België.

De Planque, P.A. 1926. *Valcooch's Regel der Duytsche Schoolmeesters. Bijdrage tot de kennis van het schoolwezen in de zestiende eeuw*. Groningen: P. Noordhoff.

De Smet, A. 1975–1976. De plaats van Antwerpen in de geografische exploratie en de cartografie der zestiende eeuw. *Tijdschrift van het Koninklijk Aardrijkskundig Genootschap van Antwerpen* 86: 4–30.

De Vries, J. 1984. *European urbanization 1500–1800*. London: Methuen & Co.

Denucé, J. 1912–1913. *Oud-Nederlandsche Kaartmakers in betrekking met Plantijn*. Antwerpen: Maatschappij der Antwerpsche Bibliophielen.

Denucé, J. 1927. *Museum Plantin Moretus, Catalogus der handschriften*. Antwerpen; Veritas.

Denucé, J. 1937. *Afrika in de 16de eeuw en de handel van Antwerpen*. Antwerpen: De Sikkel.

Denucé, J. 1941. *Koopmansleerboeken van de XVIde en XVIIde eeuwen in handschrift*.

Denucé, J. s.d. *Italiaansche koopmansgeslachten te Antwerpen in de XVIde-XVIIde eeuw*. Mechelen/Amsterdam: Het Kompas-De Spieghel.

Denucé, J., and D. Gernez. 1936. *Het zeeboek: handschrift van de Stedelijke Boekerij te Antwerpen (Nr B.29166), volume 1 of Akademie der Marine van België*. Antwerpen: De Sikkel.

Depuydt, J. 1998. De brede kring van vrienden en correspondenten rond Abraham Ortelius. In *Abraham Ortelius (1527–1598), cartograaf en humanist*, ed. P. Cockshaw and F. de Nave, 117–140. Turnhout: Brepols.

Destombes, Marcel. 1969. Deux astrolabes nautiques inédits de J. et A. de Goes, Lisbonne, 1608, 1648. *Revista de Universidade de Coimbra* 24: 3–11.

Devisscher, H. (ed.). 1995. *Fiamminghi a Roma 1508–1608*. Brussel: Snoeck-Ducaju.

Deys, H.P. 1988. Oude vestingplattegronden van Groningen, Vlissingen en Groenlo in Spaans bezit. In *Jaarboek Stichting Menno van Coehoorn*, 15–26 Utrecht: Stichting Menno van Coehoorn.

D'Hollander, R. 1993. L'astrolabe. Toulouse: Musée Paul Dupuy.

Dhombres, J., and P. Radelet-de Grave. 2008. *Une mécanique donnée à voir: les thèses illustrées défendues à Louvain en juillet 1624 par Grégoire de Saint-Vincent S.J., volume 82 (N.S. 45) of De diversis artibus*. Turnhout: Brepols.

Dibbets, G.R.W. 1983. Peeter Heyns' Cort Onderwys; Een schoolboek voor het onderwijs in de Franse taal uit de tweede helft van de zestiende eeuw. *Tijdschrift voor Nederlandse Taal- en Letterkunde* 99: 85–107.

Dibbets, G.R.W. 1994. Peeter Heyns: 'Een ghespraecksaem man, van goede gheleertheydt'. *Berichten van het Peeter Heyns Genootschap* 1: 3–13.

Dijksterhuis, E.J. 1970. *Simon Stevin: Science in the Netherlands around 1600*. Den Haag: Martinus Nijhoff.

Drake, S. 1976. Galileo and the first mechanical computing device. *Scientific American* 234(4): 104–113.

Drake, S. 1978a. *Galileo at work, his scientific biography*. Chicago/London: University of Chicago Press.

Drake, S. 1978b. *Galileo Galileï: Operations of the geometric and military compass 1606*. Washington, DC: Dibner Libraru Smithsonian Institution Press.

Duffy, C. 1996. *Siege warfare: The fortress in the early modern world 1494–1660*. London: Routledge and Kegan Paul.

Duque de Alba, don F. 1952. *Epistolaria del III Duque de Alba. Don Fernando Alvarez de Toledo*, 3 vols. Madrid: Diana.

Duverger, E. 1980–1992. *Zeventiende-eeuwse Antwerpse kunstinventarissen*, vols. 1–6. Brussel: K.V.A.W.L.S.K. van België.

Duvosquel, J. 2009. Le tarif du tonlieu et du winage de Comines établi en 1354 et son renouvellement en 1542. *Handelingen van de Koninklijke Commissie voor Geschiedenis* 175: 341–370.

Elaut, L. 1960. Pieter Hasschaerts' syfilisboekje (1564): pleidooi voor een diakritische behandeling van de venerische ziekten. *Scientiarum Historia* 2: 173–181.

Fabri, R. 1991. *De zeventiende eeuwse kunstkast, typologische en historische aspecten*. Brussel: Koninklijke Vlaamse Academie van België.

Fabri, R. 1993. *De zeventiende eeuwse kunstkast, kunsthistorische aspecten*. Brussel: Koninklijke Vlaamse Academie van België.

Favaro, A. 1895–1896. Amici e correspondenti di Galileo Galileï: Ottavio Pisani. *Atti del reale Instituto Veneto di Scienze, Lettere e Arti* 7: 411–440.

Feeman, T.G. 2002. *Portraits of the Earth; a mathematician looks at maps*. Providence: American Mathematical Society.

Folkerts, M. 1974. Die Entwicklung und Bedeutung der Visierkunst als Beispiel der praktischen Mathematik der frühen Neuzeit. *Humanismus und Technik* 18(1): 1–41.

Folkerts, M. 1977. Regiomontanus als Mathematiker. *Centaurus* 21: 214–245.

Folkerts, M. 1996. Johannes Regiomontanus – Algebraiker und Begründer der algebraischen Symbolik. In *Zur Wirkungsgeschichte der Brotordnung von Adam Ries, volume 18 of Schriften des Adam-Ries-Bundes Annaberg Buchholz*, ed. R. Gebhardt, 19–28. Annaberg-Buchholz: Adam-Ries-Bund e.V.

Folkerts, M. 2008. Die Fassmessung (Visierkunst) im späten Mittelalter und der frühen Neuzeit. In *Visier- und Rechenbücher der frühen Neuzeit, volume 19 of Schriften des Adam-Ries-Bundes Annaberg Buchholz*, ed. R. Gebhardt, 1–36. Annaberg-Burchholz: Adam-Ries-Bund e.V.

Fucikova, E. 1988a. Die Malerei am Hofe Rudolfs II. In *Prag um 1600 I*, 177–208. Freren: Luca Verlag.

Fucikova, E. 1988b. Zur Konzeption der rudolfinischen Sammlungen. In *Prag um 1600 II: Beiträge zur Kunst und Kultur am Hofe Rudolfs II*, 59–69. Freren: Luca Verlag.

Gascon, R. 1971. *Grand commerce et vie urbaine au 16e siècle: Lyon et ses marchands (environs de 1520-environs de 1580), volume 22 of Civilisations et sociétés*. Paris: Mouton.

Gatti, H. 2000. The natural philosophy of Thomas Harriot. In *Thomas Harriot: An Elizabethan man of science*, 64–92. Aldershot: Ashgate.

Gebhardt, R. (ed.). 2002. *Verfasser und Herausgeber mathematischer Texte der frühen Neuzeit, volume 14 of Schriften des Adam-Ries-Bundes Annaberg Buchholz*. Annaberg-Buchholz: Adam-Ries-Bund e.V.

Gebhardt, R. (ed.). 2005. *Arithmetische unde algebraische Schriften der frühen Neuzeit, volume 17 of Schriften des Adam-Ries-Bundes Annaberg Buchholz*. Annaberg-Buchholz: Adam-Ries-Bund e.V.

Gebhardt, R. (ed.). 2008. *Visier- und Rechenbücher der frühen Neuzeit, volume 19 of Schriften des Adam-Ries-Bundes Annaberg Buchholz*. Annaberg-Burchholz: Adam-Ries-Bund e.V.

Gebhardt, R. (ed.). 2011. *Kaufmanns-Rechenbücher und mathematische Schriften der frühen Neuzeit, volume 22 of Schriften des Adam-Ries-Bundes Annaberg-Buchholz*. Annaberg-Buchholz: Adam-Ries-Bund e.V.

Gebhardt, R., and H. Albrecht (eds.), 1996. Rechenmeister und Cossisten der frühen Neuzeit, *Freiberger Forschungshefte* 201, Freiberg.

Genard, P. 1877a. Tijdorderlijke tafel der namen van personen te Antwerpen in de XVIde eeuw voor het "feit van religie" gerechterlijk vervolgd. *Antwerpsch Archievenblad* 14: 1–103.

Genard, P. 1877b. *Naamlijst der gouverneurs van het kasteel van Antwerpen 1567–1794*. Antwerpen: De Backer.

Genard, P. 1886. *Anvers à travers les âges*. Brussel: Bruylant.

Gerlo, A. 1988. De onuitgegeven briefwisseling van Marnix van Sint-Aldegonde. Supplement III. *Mededelingen van de Koninklijke Academie voor Wetenschappen, Letteren en Schone Kunsten van België: Klasse der Letteren* 50: 43–53.

Gessner, S. 2011. The conception of a mathematical instrument and its distance from the material world. *Studium* 4: 210–227.

Gielens, A. 1933. Brieven en bescheiden van en over Willem van Oranje. *Antwerpsch Archievenblad* 38: 217–239.

Gingerich, O., and R.S. Westman. 1988. *The Wittich connection: conflict and priority in late sixteenth-century cosmology*. Philadelphia: American Philosophical Society.

Gingerich, O., and R.S. Westman. 1988. *The Wittich connection: conflict and priority in late sixteenth-century cosmology*. Philadelphia: American Philosophical Society.

Goris, J.A. 1925. *Étude sur les colonies marchandes méridionales (Portugais, Espagnols, Italiens) à Anvers de 1488 à 1567*. Leuven: Uystpruyst.

Graff, H.J. 1983. On literacy in the Renaissance: review and reflections. *History of Education* 12: 69–85.

Guicciardini, L., and C. Kilianus. 1979. *Antwerpen, Mechelen, Lier e.a.* Deurne: Soethoudt.

Gunther, R.T. 1976. *Astrolabes of the world*. London: The Holland Press.

Hall, A.R. 1969. *Ballistics in the seventeenth century*. New York: J.&J. Harper.

Haller, R. 2008. Zur Vita des Anton Neudörfer. In *Visier- und Rechenbücher der frühen Neuzeit, volume 19 of Schriften des Adam-Ries-Bundes Annaberg Buchholz*, ed. R. Gebhardt, 103–120. Annaberg-Burchholz: Adam-Ries-Bund e.V.

Haller, R. Valentin Mennher, sein Leben und seine Werke. In *Kaufmanns-Rechenbücher und mathematische Schriften der frühen Neuzeit, volume 22 of Schriften des Adam-Ries-Bundes Annaberg-Buchholz*, ed. R. Gebhardt. Annaberg-Buchholz: Adam-Ries-Bund e.V.

Haller, R. 2011. Valentin Mennher, sein Leben und seine Werke. In *Kaufmanns-Rechenbücher und mathematische Schriften der frühen Neuzeit*, ed R. Gebhardt, volume 22 of Schriften des Adam-Ries-Bundes Annaberg-Buchholz, Annaberg-Buchholz. Adam-Ries-Bund e.V. Halleux, R., and A.C. Bernès. La Cour savante d'Ernest de Bavière. Archives Internationales d'Histoire des Sciences, 5:3–29, 1995.

Hamilton, A. 1981. *The family of love*. Cambridge: Clarke.

Hamilton, A. 1987. The family of love in Antwerp. *Bijdragen tot de Geschiedenis* 70: 6–23.

Hawlitschek, K. 2005. Rechenmeister und Mathematiker im Spiegel der Ulmer Ratsprotokolle 1501–1651. In *Arithmetische und algebraische Schriften der frühen Neuzeit*, ed. R. Gebhardt, 213–224. Annaberg-Burchholz: Adam-Ries-Bund e.V.

Heeffer, A. 2005. The rhetoric of problems in algebra textbooks from Pacioli to Euler. pdf, http://logica.ugent.be/albrecht/thesis/AlgebraRhetoric.pdf. Rretrieved 27 Aug 2011.

Heeffer, A. 2006a. Van precepten tot vergelijkingen: de conceptuele ontwikkeling van de symbolische algebra in de zestiende eeuw. Ph.D. thesis, Universiteit Gent.

Heeffer, A. 2006b. The origin of the problems in Euler's Algebra. *Bulletin of the Belgian Mathematical Society – Simon Stevin* 5: 949–952.

Heijting, W. 2007. Boeken en lectuur in het Behouden Huys. In *Profijtelijke boekskens: boekcultuur, geloof en gewin: historische studies*, 147–166. Hilversum: Verloren.

Imhof, D. (ed.), 1994. Gerard Mercator en de geografie in de Zuidelijke Nederlanden, Antwerpen: Publikaties van het Museum Plantin Moretus en het Stedelijk Prentenkabinet 29.

Imhof, D. (ed.). 1998. *Tentoonstellingscatalogus Abraham Ortelius*. Antwerpen: Publikaties van het Museum Plantin Moretus en het Stedelijk Prentenkabinet.

Israel, J.I. 1989. Dutch primacy in world trade 1585–1740. Oxford: Clarendon Press.

Israel, J.I. 1998. *The Dutch Republic*. Oxford history of early modern Europe. Oxford: Clarendon Press.

Itard, J. 1976. Denis Henrion. In *Dictionary of scientific biography VI*, 271. Washington, DC: Dibner's.

Jaeger, W. 1976. *Die Illustrationen von Peter Paul Rubens zum Lehrbuch der Optik des Franciscus Aguilonius, 1613*. Heidelberg: Brausdruck.

Janssens, G. 1976. Het oordeel van tijdgenoten en historici over Alva's bestuur in de Nederlanden. *Belgisch Tijdschrift voor Filologie en Geschiedenis* 54: 474–488.

Janssens, G. 1981. Brabant in het verweer: Absolute monarchie of staatsbewind van Alva tot Farnese, 1567–1578. Ph.D. thesis, Katholieke Universiteit Leuven.

Janssens, G. 2009. Het "politiek testament" van de hertog van Alva. *Handelingen van de Koninklijke Commissie voor Geschiedenis* 175: 447–474.

Johnston, S. 1994. Making mathematical practice; gentlemen, practitioners and artisans in Elizabethan England. PhD thesis, University of Cambridge.

Jonkers, A. 2008. The pursuit of magnetic shadows: The formal-empirical dipole field of early-modern geomagnetism. *Centaurus* 50: 254–289.

Jorink, E. 1994. Hemelse tekenen. Nederlandse opvattingen over de komeet van 1618. *Gewina* 17(2): 68–82.

Kaunzner, W. 2005. Ein Weg zu den Rechenmeister im fränkischen Raum. In *Arithmetische unde algebraische Schriften der frühen Neuzeit, volume 17 of Schriften des Adam-Ries-Bundes Annaberg Buchholz*, ed. R. Gebhardt, 171–198. Annaberg-Buchholz: Adam-Ries-Bund e.V.

Kennerley, A., and P. Seymour. 2000. Aids to the teaching of nautical astronomy and its history from 1600. *Pædagogica Historica* 36: 151–175.

Keuning, J. 1950. Cornelis Anthonisz. *Imago Mundi* 7: 51–65.

Keuning, J. 1956. The Van Langren family. *Imago Mundi* 13: 101–109.

Kinget, J. 1989. De familie Kinget in Antwerpen. *De Cygnet* 1: 25–26.

Kline, M. 1972. *Mathematical thought from ancient to modern times*. New York: Oxford University Press.

Koeman, C. 1994. *Gewestkaarten van de Nederlanden door Jacob van Deventer, 1536–1545: met een picturale weergave van alle kerken en kloosters*. Alphen aan den Rijn: Canaletto.

Koeman, C., and J.C. Visser. 1992–. *De stadsplattegronden van Jacob van Deventer*. Landsmeer: Robas BV.

Kool, M. 1999. *Die conste vanden getale*. Uitgeverij Verloren, Hilversum.

Kronenberg, M.E. 1928. Een onbekende almanack van Dodonaeus (1540). *Het Boek* 17: 191–194.

Kubler, G. 1964. Francesco Pacciotto, architect. In *Essays in the memory of Karl Lehman*, 176–189. New York: Institute of Fine Arts, New York University.

Lach, D.F. 1977. *Asia in the making of Europe: Volume II, a century of wonder*. Chicago: University of Chicago Press.

Lamb, U. 1972. *A navigator's universe: The 'Libro de Cosmographia' of 1538 by Pedro de Medina*. Chicago/London: University of Chicago Press.

Lambour, R.C. 2004. 400 jaar volgens Bartjens: Meester Willem Bartjens (1569–1638) en de zijnen in Amsterdam en Zwolle, zijn schoonfamilie Stangerus in Amsterdam en Delft en de Delftse regent Andries van Stangerlant als erflater. *De Nederlandsche Leeuw* 121: 76–112.

Laurent, R. 1986. *De havens aan de kust en aan het Zwin*. Brussel: Algemeen Rijksarchief en Rijksarchief in de Provinciën.

Lebègue, R. 1943. *Les correspondants de Peiresc dans les anciens Pays-Bas*. Brussel: Lebègue.

Lefrancq, J. 2008. Hémisphère nautique. In *Séminaire de Tournai; Histoire-Bâtiments-Collections*, ed. Monique Maillard-Luypaert, 126–127.

Leibowitz, G. 1933. Die Visierkunst im Mittelalter. Ph.D. thesis, Universität zu Heidelberg.

Leitão, H. 2007. Maritime discoveries and the discovery of science: Pedro Nunes and early modern science. In *Mas allá de la leyenda negra: España y la revolución científica; Beyond the black legend: Spain and the scientific revolution*, ed. V. Navarro Brotóns and W. Eamon, 89–104, Valencia. Universitat de València.

Lesger, C. 2001. *Handel in Amsterdam ten tijde van de Opstand*. Hilversum: Verloren.

Lloyd, T.H. 1982. *Alien merchants in England in the high middle ages*. Sussex: Harvester.

Lombaerde, P. 2009. *Antwerpen versterkt: de Spaanse omwalling vanaf haar bouw in 1542 tot haar afbraak in 1870*. Brussel: UPA.

Maclean, I. 1990. Philosophical books in European markets 1570–1630: The case of Ramus. In *New perspectives on Renaissance thought: Essays in the history of science, education and philosophy*, ed. J. Henry and S. Hutton, 253–263. London: Duckworth.

Maor, E. 1998. *Trigonometric delights*. Princeton: Princeton University Press.

Marinus, M.J. 1989. *Laevinus Torrentius als tweede bisschop van Antwerpen (1587–1595)*. Brussel: K.A.W.L.S.K.

Marnef, G. 1992. Repressie en censuur in het Antwerps boekbedrijf 1567–1576. *De Zeventiende Eeuw* 8: 221–231.

Marnef, G. 1994. *Prosopografie van protestanten en rebellen te Antwerpen vervolgd 1550–1577*. Antwerp: UA treatise.

Marnef, G. 1996a. *Antwerp in the age of reformation: Underground protestantism in a commercial metropolis*. Baltimore: Johns Hopkins University Press.

Marnef, G. 1996b. *Antwerpen in de tijd van de Reformatie: ondergronds protestantisme in een handelsmetropool 1550–1577*. Antwerpen: Kritak.

Marnef, G. s.d. De boedelbeschrijving van W. Sylvius. unpublished ca. 1994.

Martens, P. 2009. Militaire architectuur en vestingoorlog tijdens het regentschap van Maria van Hongarije (1531–1555). De ontwikkeling van de gebastioneerde verstingbouw. Ph.D. thesis, Katholieke Universiteit Leuven, Leuven.

Masure, T. 1986. De stadsfinanciën van Antwerpen 1531–1571. een poging tot reconstructie. Master's thesis, Rijksuniversiteit Gent.

Meeus, H. 1988–1989. Zacharias Heyns, een leerjongen van Jan Moretus. *De Gulden Passer* 67: 599–611.

Meeus, H. 2000. Peeter Heyns, a "French Schoolmaster". In *Grammaire et enseignement du français 1500–1700*, 301–316. Leuven: Peeters.

Meeus, H. 2003. Peeter Heyns' "Le miroir des vefves" meer dan schooltoneel?. In *Mémoire en temps advenir*, 115–134. Leuven: Peeters.

Meeus, H. 2009. De Spieghel der werelt als spiegel van Peter Heyns. In *Ortelius' Spieghel der Werelt*. Antwerpen: Vereniging van Antwerpse Bibliofielen.

Meskens, A. 1992. Michiel Coignet's nautical instruction. *The Mariner's Mirror* 78: 257–276.

Meskens, A. 1993a. Zestiende-eeuwse wiskundige sprokkelingen vnl. uit het Antwerps stadsarchief. *Scientiarum Historia* 19: 23–28.

Meskens, A. 1993b. Een zestiende-eeuws Antwerps rekenschrift. *Nieuw Tijdschrift van de Vrije Universiteit Brussel* 6: 109–114.

Meskens, A. 1994a. Winegauging in late sixteenth and early seventeenth Century Antwerp. *Historia Mathematica* 21: 121–147.

Meskens, A. 1994b. *Wiskunde tussen Renaissance en Barok, aspecten van wiskunde-beoefening te Antwerpen 1550–1620, volume 41–43 of Publikaties van de Stadsbibliotheek*. Antwerpen: Stadsbibliotheek Antwerpen.
Meskens, A. 1994c. Zeevaartkunde ten tijde van Mercator: enkele aspecten *Gerard Mercator en de geografie in de Zuidelijke Nederlanden*, 42–50. Museum Plantin Moretus en Stedelijk Prentenkabinet.
Meskens, A. 1995. De prijs van de wetenschap: enkele prijzen van wiskundige en andere wetenschappelijke boeken in Plantijnse archivalia. *De Gulden passer* 73: 83–106.
Meskens, A. 1996a. Mathematics Education in late sixteenth Century Antwerp. *Annals of Science* 53: 137–155.
Meskens, A. 1996b. Peter Heyns. In *Nationaal Biografisch Woordenboek 15*, 338–342. Brussel: Koninklijke Vlaamse Academie van België.
Meskens, A. 1996c. Enkele biografische gegevens over Gillis i Coignet alias Gillis met de Vlek. *Oud Holland* 110: 142–145.
Meskens, A. 1997a. Michiel Coignet's contribution to the development of the sector. *Annals of Science* 54: 143–160.
Meskens, A. 1997b. De doos van Pandora. een kaart van Michiel Coignet. *Gewina* 20: 32–34.
Meskens, A. 1997c. The Jesuit Mathematics School in Antwerp in the early seventeenth Century. *The Seventeenth Century* 12: 11–22.
Meskens, A. 1998a. Familia universalis: Coignet. Antwerpen: Koninklijk Museum voor Schone Kunsten.
Meskens, A. 1998b. De wereld op het platte vlak: wiskundige cartografie ten tijde van Abraham Ortelius. In *Abraham Ortelius*, 71–82. Turnhout: Brepols.
Meskens, A. 1998–1999. Liaisons dangereuses: Peter Heyns en Abraham Ortelius. *De Gulden Passer* 76–77: 96–108.
Meskens, A. 1999. The Portrait of Jan-Karel della Faille by Anthony Van Dijck. *Koninklijk Museum voor Schone Kunsten Jaarboek* 39: 124–137.
Meskens, A. 2001. Een familie herenigd met haar instrument. *Scientiarum Historia* 27(1): 73–82.
Meskens, A. 2002. Some new biographical data about Michiel Coignet. *Nuncius* 17: 447–454.
Meskens, A. 2005a. *Joannes della Faille s.j.* Brussel/Rome: Belgisch Historisch Instituut te Rome.
Meskens, A. 2005b. Abdijen en een stukje vergeten wiskundegeschiedenis. In *Abdijbibliotheken, verleden, heden, toekomst*, 123–133. Antwerpen: Vereniging van Antwerpse Bibliofielen.
Meskens, A., G. Bonte, J. De Groote, M. De Jonghe, and D.A. King. 1999. Wine gauging at Damme: The evidence of a late medieval manuscript. *Histoire et mesure* 14: 51–77.
Meurer, P.H. 1991. *Fontes cartographici Orteliani: das "Theatrum orbis terrarum" von Abraham Ortelius und seine Kartenquellen*. Weinheim: VCH, Acta Humaniora.
Michel, H. 1947. *Traité de l'Astrolabe*. Paris: Gauthier-Villars.
Müller, J. 1959. Les ingénieurs dans les Pays-Bas espagnols (1500–1715). *Revue Internationale d'Histoire Militaire* 33: 467–478.
Mosley, A. 2007. *Bearing the heavens*. Cambridge: Cambridge University Press.
Mörzer-Bruyns, W.F.J. 1984. Leeskaarten en paskaarten uit de Nederlanden. In *Lucas Jansz. Waghenaer van Enckhuysen*, 11–20. Enkhuizen: Vrienden van het Zuiderzeemuseum.
Mörzer-Bruyns, W.F.J. 1994. *The cross staff. History and development of a navigational instrument*. Amsterdam: VNHSM.
Nauwelaerts, M.A. 1978. Drukkers en schoolboeken te Antwerpen tot 1600. *Varia Historica Brabantia* 6–7: 273–300.
North, J.D. 1974. The astrolabe. *Scientific American* 230(1): 96–106.
Page, W. 1893. *Letters of Denization and acts of Naturalisation for Aliens in England 1509–1603*. Lymington: The Huguenot Society.
Parker, G. 1972. *The army of Flanders and the Spanish road*. Cambridge: Cambridge University Press.
Parker, G. 1980. An educational revolution? The growth of literacy and schooling in modern Europe. *Tijdschrift voor Geschiedenis* 93: 210–220.
Parker, G. 1990. *Spain and the Netherlands. Ten Studies*. Glasgow: Fontana Press.

Parker, G. 2004. *The army of Flanders and the Spanish road*, 2nd ed. Cambridge: Cambridge University Press.

Pauly, M. 1994. *Luxemburg im späten Mittelalter 2: Weinhandel und Weinconsum*. Luxembourg: CLUDEM.

Penneman, T., and A. van der Gucht. 1994. *Mercator en zijn boeken*. St-Niklaas: Koninklijke Oudheidkundige Kring van het Land van Waas.

Pinchart, A. 1860. *Archives des Arts, Sciences et Lettres*, 2 vols. Gent: Hebbelynck.

Poffé, E. 1895. De gilde van de Antwerpsche schoolmeesters van bij haar ontstaan tot haar afschaffing. Antwerpen: Kennes.

Pogo, A. 1935. Gemma Frisius, his method of determining longitude. *ISIS* 22: 469–485.

Portet, P. 1993–1994. Le Systeme métrologique de Paris au Moyen Age. *Cahiers de métrologie* 11–12: 463–488.

Prims, F. 1929. Het huis van Peter Heyns in de Muntstraat. *Antwerpiensia* 3: 129–136.

Prims, F. 1931. Letterkundigen, geleerden en kunstenaars. Verslagen en Mededeelingen van de Koninklijke Vlaamsche Academie voor Taal- en Letterkunde, maart: 171–221.

Prims, F. 1932. Gillis Hooftman. *De zondagsvriend* 21: 521.

Prims, F. 1933. De Herentalsche Vaart. *Antwerpiensia* 6: 405–406.

Prims, F. 1938. Het altaar van de kolveniers. *Antwerpiensia* 12: 302–308.

Prims, F. 1941. De scholen onder de regiemen van den religionsvrede en het Calvinisme. *Antwerpiensia* 15: 263–272.

Prims, F. 1948a. Michiel Coignet. *Antwerpiensia* 19: 103–114.

Prims, F. 1948. Michiel Coignet. *Antwerpiensia* 19: 103–114.

Prims, F. 1951a. De scholen te Antwerpen onder de religionsvrede 1578–1581. *Bijdragen tot de Geschiedenis* 3: 212–226.

Puraye, J. 1968. Abraham Ortelius, Album amicorum: notes et traduction. *De gulden passer/Vereeniging der Antwerpsche Bibliophielen* 46: 1–99.

Put, E. 1990. *De Cleijne Scholen. Het volksonderwijs in het hertogdom Brabant tussen katholieke Reformatie en Verlichting (eind 16de eeuw-1795)*. Leuven: Universitaire Pers.

Rasquin, V.A. 1997a. Atribuido a M. Coignet y F. Arsenius: teodolito simple. In *Instrumentos cientificos del siglo XVI, La corte Española y la escuela de Lovaina*, ed. J. Van Damme and K. van Cleempoel, 228–229. Madrid: Fundacion Carlos de Amberes.

Rasquin, V.A. 1986–1988. Les règles à calcul anciennes et leur utilisation en navigation: les compas de proportion, les règles de Gunther. *Mededelingen van de Marine Academie* 28: 53–87.

Rasquin, V.A. 1996. *Dictionnaire des constructeurs belges d'instruments scientifiques*. Bruxelles: Comité Nationale de Logique, d'Histoire et de Philosophie des Sciences.

Reich, U. 2005. Über hundert Auflagen: das Arithmetikbuch des Gemma Frisius (1508–1555). In *Arithmetische unde algebraische Schriften der frühen Neuzeit, volume 17 of Schriften des Adam-Ries-Bundes Annaberg Buchholz*, ed. R. Gebhardt. Annaberg-Buchholz: Adam-Ries-Bund e.V.

Rekers, B. 1972. *Benito Arias Montano*. London: Warburg Institute.

Renard, A. 1921. *La tonnelerie à la portée de tous*. Paris: E. Malfère.

Rochhaus, P. 1996. Adam Ries und die Annaberger Rechenmeister zwischen 1500 und 1604, in *Rechenmeister und Cossisten der frühen Neuzeit*, ed. Gebhardt, R., and H. Albrecht (eds.), Freiberger Forschungshefte 201, Freiberg.

Roegiers, J., and B. vander Herten (eds.). 1994. *Eenheid op papier*. Leuven: Davidsfonds.

Rombouts, P., and T. van Lerius. 1874. *De Liggeren en andere historische archieven der Antwerpsche Sint-Lucasgilde*. Antwerpen: F. Baggerman.

Rome, A. 1927. L'astrolabe et le météoroscope. *Annales de la Societé Scientifique de Bruxelles, Mémoires* 47: 77–102.

Roosens, B.H. 2005. Habsburgse defensiepolitiek en vestingbouw in de Nederlanden (1520–1560). Ph.D. thesis, Leiden University.

Rooses, M., and J. Denucé. 1968. *Correspondance de Christophe Plantin*, 10 vols. Nendeln: Klaus Reprint.

Rose, P.L. 1968. The origins of the proportional compass from Mordente to Galileo. *Physis* 10: 53–69.

Röttel, K. 1996. Der Beitrag des Peter Apian zur Mathematik, Astronomie, Geographie und Physik. In *Rechenmeister und Cossisten der frühen Neuzeit, volume 201 of Freiberger Forschungshefte*, ed. R. Gebhardt and H. Albrecht, 19–28. Freiberg: Technische Universität Bergakademie.

Rouzet, A. 1975. *Dictionnaire des imprimeurs, libraires et éditeurs belges de XVe et XVIe siècle*. Den Haag: Nieuwkoop-De Graaf.

Rüdiger, B. 2005. Gedruckte Rechenbücher im Besitz von Leipziger Einwohnern bis 1600. In *Arithmetische und algebraische Schriften der frühen Neuzeit*, ed. R. Gebhardt, 421–434. Annaberg-Buchholz: Adam-Ries-Bund e.V.

Sabbe, M. s.d. *Peeter Heyns en de Nimfen uit den Lauwerboom*. Antwerpen/Den Haag: De Coker-Nijhoff.

Schlugleit, D. 1936. Alphabetische naamlijst op de Goud- en Zilversmeden te Antwerpen voor 1600. *Bijdragen tot de Geschiedenis* 27: 6–69.

Schlugleit, D. 1969. De Antwerpse goud-en zilversmeden in het corporatief stelsel (1332–1798). Wetteren: Bracke.

Schneider, I. 1970. Der Proporzionalzirkel, ein universelles Analogrecheninstrument der Vergangenheit.

Schneider, I. 2002. Ausbildung und fachliche Kontrolle der deutschen Rechenmeister vor dem Hintergrund ihrer Herkunft und ihres sozialen Status. In *Verfasser und Herausgeber mathematischer Texte der frühen Neuzeit, volume 14 of Schriften des Adam-Ries-Bundes Annaberg Buchholz*, ed. R. Gebhardt, 103–112. Annaberg-Buchholz: Adam-Ries-Bund e.V.

Schoffer, I., H. van der Wee, and J.A. Bornewasser. 1988. *De Lage Landen 1500–1780*. Amsterdam: Agon.

Schofield, C.J. 1961. *Tychonic and semi-Tychonic world systems*. New York: Arno Press.

Schoups, I., and A. Wiggers (eds.). 1998. *Philips van Marnix van Sint Aldegonde*. Antwerpen: Pandora.

Schröder, E. Ulrich Wagner, Autor des ersten gedruckten deutschsprachigen kaufmännischen Rechenbuches von 1483, 29–36.

Serrure, C.P. 1859–1860. *Vaderlandsch Museum III*. Gent: H. Hoste.

Shen, Kangren et al. (Trans.). 1999. *The nine chapters on the mathematical art: Companion and commentary*. Oxford: Oxford University Press.

Sillem, W. 1883. Zur Geschichte der Niederländer in Hamburg von ihrer Ankunft bis zum Anschluss des Niederländischen Contracts 1605. *Zeitschrift der hamburgische Geschichte* 7: 481–598.

Sillem, W. 1893. Die wallonische Gemeinde in Stade. *Geschichtsblätter des Deutschen Hugenotten Vereins II* 6: 3–32.

Simi, A., and L. Toti-Rigatelli. 1993. Some 14th and 15th Century Texts on Practical Geometry. In *Vestigia Mathematica*, ed. M. Folkerts and J.P. Hogendijk, 453–470. Amsterdam: RODOPI.

Sitters, M.H. 2008. *Sybrandt Hansz Cardinael; rekenmeester en wiskundige*. Hilversum: Verloren.

Smeur, A.J.E.M. 1960a. *De zestiende-eeuwse Nederlandse rekenboeken*. Den Haag: Martinus-Nijhoff.

Smeur, A.J.E.M. 1960b. Bestaat er nog een exemplaar van de eerste uitgave van de arithmetica van Willem Raets?. *Scientiarum Historia* 2: 22–23.

Smith, D.E. 1959. *A source book in mathematics*. New York: Dover Publ.

s.n. 1863. *Antwerpsche parochiekerken, volume 2 of Verzameling der graf- en gedenkschriften van de provincie Antwerpen*. Antwerpen: J.E. Buschman.

s.n. 1864. De rekening van de stad antwerpen 1549–1550. *Antwerpsch Archievenblad* 1: 17–104.

s.n. 1964. *Het beleg van Hulst*. Utrecht: Stichting Menno van Coehorrn.

s.n. 1985. *The Planispheric astrolabe*. National Maritime Museum: Greenwich.

Sobel, D., and W.J.H. Andrewes. 1998. *The illustrated longitude*. New York: Walker & Co.

Soen, V. 2007. *Geen pardon zonder paus!: Studie over de complementariteit van het koninklijk en pauselijk generaal pardon (1570–1574) en over inquisiteur-generaal Michael Baius (1560–1576)*. Brussel: KVAB.

Soly, H. 1968. De brouwerijenonderneming van Gilbert van Schoonbeke (1552–1562). *Belgisch Tijdschrift voor Filologie en Geschiedneis* 46: 337–392 and 1166–1204.

Soly, H. 1974. De schepenregisters als bron voor de conjunctuurgeschiedenis van Zuid- en Noordnederlandse steden in het Ancien Regime. een concreet voorbeeld: de Antwerpse immobiliënmarkt in de zestiende eeuw. *Tijdschrift voor Geschiedenis* 87: 521–544.

Soly, H. 1976. De bouw van de Antwerpse citadel (1567–1571). Sociaaleconomische aspecten. *Belgisch Tijdschrift voor Militaire Geschiedenis* 21: 548–578.

Soly, H. 1977. *Urbanisme en kapitalisme te Antwerpen in de 16de eeuw: de stedebouwkundige en industriële ondernemingen van Gilbert van Schoonbeke*. Brussels: Gemeentekrediet van België.

Sosson, J.P. 1990. Wijn en gemeentelijke inkomsten. In *Van Rank tot Drank*, ed. E. Van Driessche, 200–202.

Stimson, A. 1988. *The mariner's astrolabe*. Utrecht: H&S Publishers.

Struik, D.J. 1980. *Geschiedenis van de wiskunde*. Amsterdam: SUA.

Struik, D.J. 1980. *Geschiedenis van de wiskunde*. Amsterdam: SUA.

Subacchi, P. 1995. Italians in Antwerp in the second half of the 16th century. In *Minderheden in Westeuropese steden (16de-20ste eeuw)*, ed. H. Soly and A.K.L. Thijs, 73–90. Rome: Academia Belgica.

Swart, K.W. 1984. Wat bewoog Willem van Oranje de strijd tegen de Spaanse overheersing aan te binden?. *Bijdragen en mededelingen betreffende de geschiedenis der Nederlanden* 99: 554–572.

Swerdlow, H.M., and O. Neugebauer. 1984. *Mathematical astronomy in Copernicus' De Revolutionibus*. New York: Springer.

Swiggers, P., and J. De Clercq. 1995. Franse grammatica en taalonderwijs in de "Lage Landen" in de zestiende en zeventiende eeuw. *Meesterwerk. Berichten van het Peeter Heyns Genootschap* 2: 25–35.

ter Kuile, O. 1986. *Koper en Brons*. Amsterdam: Rijksmuseum.

Terpstra, H. 1950. *Jacob van Neck, Amsterdams admiraal en regent*. Amsterdam: Van Kampen.

Thijs, A.K.L. 1987. De Contrareformatie en het economisch transformatieproces te Antwerpen na 1585. *Bijdragen tot de Geschiedenis* 70: 97–124.

Thijs, G. 1980. *Rembert Dodoens, zijn leven, zijn tijd, zijn werken*. Delft: SPD Rijperman.

Thomas, K. 1989. *De ondergang van de magische wereld. Godsdienst en magie in Engeland 1500–1700*. Amsterdam: Agon.

Thomas, W. 2004a. Het beleg van Oostende. In *De val van het nieuwe Troje; Het beleg van Oostende, 1601–1604*, 81–100. Leuven: Davidsfonds.

Thomas, W. (ed.). 2004b. *De val van het nieuwe Troje; Het beleg van Oostende, 1601–1604*. Leuven-Oostende: Davidsfonds.

Thomas, W. 2004c. Het beleg van Oostende. In *De val van het nieuwe Troje; Het beleg van Oostende, 1601–1604*, ed. W. Thomas, 81–100. Leuven/Oostende: Davidsfonds.

Thoren, V.E. 1990. *The lord of Uraniborg, a biography of Tycho Brahe*. Cambridge: Cambridge University Press.

Tropfke, J. 1980. *Geschichte der Elementarmathematik: Arithmetik und Algebra*, vol. I, 4th ed. Berlin/New York: W. De Gruyter.

Turner, G.L'E. 1983. Mathematical instrument making in London in the sixteenth century. In *English map making 1500–1650*, ed. S. Tyacke, 93–106. London: British Library.

Turner, G.L'E. 2006. A novel Italian-hour nocturnal by Michiel Coignet. *Annals of Science* 63(2): 215–219.

Van Acker, J. 1982. Het memoriaalboek van Antonio II Anselmo. *Bijdragen tot de Geschiedenis* 65: 79–108.

van Cauwenberghe, M. 1986. De Stadsfinanciën in Damme. Master's thesis, Rijksuniversiteit Gent, Gent.

van Cleempoel, K. 1997a. Instrumentos cientificos en el Escorial. In *Instrumentos cientificos del siglo XVI, La corte Española y la escuela de Lovaina*, ed. J. Van Damme and K. van Cleempoel, 67–84. Madrid: Fundacion Carlos de Amberes.

van Cleempoel, K. 1997b. La produccion de instrumentos cientificos en Lovaina en el siglo XVI. In *Instrumentos cientificos del siglo XVI, La corte Española y la escuela de Lovaina*, ed. J. Van Damme and K. van Cleempoel, 87–110. Madrid: Fundacion Carlos de Amberes.
van Cleempoel, K. 2002. *A Catalogue Raisonné of Scientific instruments from the Louvain School, 1530 to 1600*. Turnhout: Brepols.
van Cleemput, F. 1973–1975. Michiel Coignet, een Antwerps wis- en zeevaartkundige uit de zestiende eeuw. *Mededelingen Marine Academie* 23: 112–148.
Van Damme, J., and K. van Cleempoel (eds.). 1997. *Instrumentos cientificos del siglo XVI, La corte Española y la escuela de Lovaina*. Madrid: Fundacion Carlos de Amberes.
van de Vijver, O. 1977. Lettres de Jean Charles della Faille s.j., cosmographe du roi à Madrid à M. Fl. van Langren, cosmographe du roi à Bruxelles. *Archivum Historicum Societatis Iesu* 46(91): 73–183.
van de Vijver, O. 1980. L'école de mathématiques des Jésuites de la province flandor-belge au xviie siècle. *Archivum Historicum Societatis Iesu* 49(97): 265–271.
van de Vijver, O. 1989. De wetenschappen. In *Antwerpen in de 17e eeuw*. Antwerpen: Genootschap voor Antwerpse Geschiedenis.
Van de Vyver, O. 1975. *Jan Ciermans (Pascasio Cosmander) 1602–1648, wiskundige en vestingbouwer, volume 7 of Mededelingen uit het Seminarie voor geschiedenis van de wiskunde en de natuurwetenschappen aan de Katholieke Universiteit te Leuven*. Leuven: Departement wiskunde, Katholieke Universiteit Leuven.
van de Weyer, D. 2011. Opera Mathematica: La Géometrie, S. Marolois. Master's thesis, Vrije Universiteit Brussel, Brussel.
van den Heuvel, C. 1991. *'Papiere Bolwercken'; De introductie van de Italiaanse stede- en vestingbouw in de Nederlanden (1540–1609) en het gebruik van tekeningen*. Alphen aan den Rijn: Canaletto.
van der Krogt, P. 1994. *Globi Neerlandici, the production of globes in the low countries*. Utrecht: HES.
van der Stock, J. 1993. *Antwerpen, verhaal van een metropool*. Antwerpen: Snoeck Ducaju & zoon.
van der Wee, H. 1987. Opkomst van een wereldstad: handel en nijverheid te Antwerpen van de veertiende tot de achttiende eeuw. *Academia Analecta* 49(2): 3–18.
van Durme, M. 1959. *Correspondance mercatorienne*. Antwerpen: DNB.
van Egmond, W. 1976. The commercial revolution and the beginnings of Western mathematics in Renaissance Florence 1300–1500. Ph.D. thesis, Indiana University.
van Eysinga, W.J.M. 1959. *De wording van het Twaalfjarig Bestand van 9 april 1609*. Amsterdam: Noord-Hollandsche Uitgeversmij.
van Foreest, H.A. (ed.). 1980–1981. *De vierde schipvaart der Nederlanders naar Oost-Indië onder Jacob Wilkens en Jacob van Neck (1599–1604), volume 81–82 of Werken uitgegeven door de Linschoten-Vereeniging*. 's-Gravenhage: Nijhoff.
van Hemeldonck, G. 1998. Repertorium der Belgische goud- en zilversmeden. unpublished.
van Looy, H. 1979. Chronologie en analyse van de mathematische handschriften van Gregorius a Sancto Vincentio. Ph.D. thesis, Katholieke Universiteit Leuven, Leuven.
van Looy, H. 1980. Chronologie et analyse des manuscrits mathématiques de Grégoire de Saint Vincent (1584–1667). *Archivum Historicum Societatis Iesu* 49: 279–303.
van Looy, H. 1984. A chronology and historical analysis of the mathematical manuscripts of Gregorius a Sancto Vincentio (1584–1667). *Historia Mathematica* 11: 57–75.
van Nouhuys, T. 1998a. *The age of two-faced Janus: The comets of 1577 and 1618 and the decline of the Aristotelean world view in the Netherlands*. Leiden: Brill.
van Nouhuys, T. 1998b. Copernicus als randverschijnsel: de kometen van 1577 en 1618 en het wereldbeeld in de Nederlanden. *Scientiarum Historia* 24: 17–38.
van Ortroy, F. ca. 1910. *L'école cartographique belge au XVIe siècle*. Louvain: s.n.
van Ortroy, F. 1914. *L'oeuvre cartographique de Gérard et de Corneille de Jode*. Gent: E. Van Goethem.

van Ortroy, F. 1920. *Bio-bibliographie de Gemma Frisius et ses neveux les Arsenius*. Brussel: Lametrin.

van Roey, J. 1963. De sociale structuur en de godsdienstige gezindheid van de Antwerpse bevolking op de vooravond van de Reconciliatie met Farnese (17 augustus 1585). Ph.D. thesis, Rijksuniversiteit Gent, Gent.

van Roey, J. 1968. De correlatie tussen het sociale beroepsmilieu en de godsdienstkeuze te Antwerpen op het einde der XVIde eeuw. In *Bronnen voor de religieuze geschiedenis van België (Middeleeuwen en Moderne Tijden)*. Leuven: s.n.

van Roey, J. 1985. De "zuivering" van de Antwerpse gewapende gilden na de val van de stad (17-8-1585). *Taxandria* 7: 199–206.

van Roosbroeck, J. 1968. *Emigranten, Nederlandse vluchtelingen in Duitsland*. Leuven: Davidsfonds.

van Valkenburg, C.C. 1988. Quignet-Coignet. *Jaarboek Centraal Bureau voor Genealogie* 32: 71–90.

Vande Winkel, G. 2008. Leven en werk van cartograaf Jacob van Deventer. *Het Land van Aalst* 60: 225–242.

Vanden Broecke, S. 2000. The use of visual media in Renaissance cosmography: The Cosmography of Peter Apian and Gemma Frisius. *Paedagogica Historica* 36: 131–150.

Vandommele, J.J.M. 2011. *Als in een spiegel. Vrede, kennis en gemeenschap op het Antwerpse Landjuweel 1561*. Hilversum: Verloren.

Verlinden, C. 1987. De Nederlandse vertaling van het 'Arte de Navegar' van Pedro de Medina en de nautische 'Onderwijsinghe' van Michiel Coignet (Antwerpen, 1580). *Collectanea Maritima* 3: 5–20.

Vermeir, R. 2001. *In staat van oorlog; Filips IV en de Zuidelijke Nederlanden 1629–1648*. Maastricht: Shaker.

Vermeir, R. 2011. De (Zuid-)Nederlandse aristocratie en de vorming van een transnationale elite in de Spaans-Habsburgse samengestelde staat. In *Werken aan de stad*, 291–310. Brussels: VUBPress.

Voet, L. 1980–1983. *The Plantin Press (1555–1589). A Bibliography of the Works printed by Christopher Plantin at Antwerp and Leiden*. Amsterdam: Van Hove.

Voet, L. 1989. Christoffel Plantijn (ca. 1520–1589), drukker van het humanisme. In *Christoffel Plantijn en de exacte wetenschappen in zijn tijd*, ed. E. Cockx-Indestege and F. De Nave, 33–43, Brussel: Gemeentekrediet van België.

Waller-Zeper, C.M. 1937. *De oudste intresttafels in Italië, Frankrijk en Nederland met een herdruk van Stevins 'tafelen van interest'*. Amsterdam: Noord-Hollandsche Uitgeversmij.

Ward, F.A.B. 1981. European scientific instruments in the Department of Medieval and later antiquities of the British Museum. London: British Museum Publications.

Waterbolk, E.H. 1974. The 'Reception' of Copernicus's teachings by Gemma Frisius (1508–1555). *Lias* 1: 225–242.

Waters, D.W. 1978. The art of navigation in England in Elizabethan and early Stuart times.

Wauwermans, H. 1894. *Les fortifications d'Anvers au XVIème siècle*. Antwerpen: De Backer.

Webb, H.J. 1945. The mathematical and military works of Thomas Digges, with an account of his Life. *Modern Language Quarterly* 6: 389–400.

Wegg, J. 1979. *The decline of Antwerp under Philip of Spain*. Westport: Hyperion Press.

Westra, F. 1992. Nederlandse ingenieurs en de fortificatiewerken in het eerste tijdperk van de Tachtigjarige oorlog, 1573–1604. Uitg. Vis, Alphen aan den Rijn.

Wijnroks, E.H. 2003. *Handel tussen Rusland en de Nederlanden 1560–1640*. Hilversum: Verloren.

Woltjer, J.J. 1994. *Tussen vrijheidsstrijd en burgeroorlog*. Amsterdam/Leuven: Balans-Kritak.

Yeomans, D.K. 1991. *Comets, a chronological history of observation; science, myth and folklore*. New York: Wiley.

Zeller, O. 1983. *Les Récensenments lyonnais de 1597 et 1636, démographie historique*. Lyon: Presses universitaires de Lyon.

Zemon-Davis, N. 1960. Sixteenth century French arithmetics on the business life. *Journal of the History of Ideas* 21: 18–48.

Ziggelaar, A. 1983. *François de Aguilon s.j. (1567–1617). Scientist and Architect, volume 44 of Bibliotheca Instituti Historici s.i.* Roma: Institutum Historicum s.i.

Ziggelaar, A. 1995. Jesuit astronomy north of the Alps. four unpublished jesuit letters 1611–1620. In *Christoph Clavius e attività scientifica dei gesuiti nell'eta di Galileo*, ed. U. Baldini, 101–132. Roma: Bulzoni Editore.

Zinner, E. 1956. *Deutsche und Niederländische astronomische Instrumente des 11. bis 18. Jahrhundert*. München: C.H. Beck'sche Verlagsbuchhandlung.

Zinner, E. 1968. *Leben und Wirke des Joh. Müller von Koningsberg genannt Regiomontanus*. Osnabrück: Otto Zeller.

Index

A. Meskens, *Practical Mathematics in a Commercial Metropolis: Mathematical Life in Late 16th Century Antwerp*, Archimedes 31, DOI 10.1007/978-94-007-5721-9,
© Springer Science+Business Media Dordrecht 2013

MIX
Papier aus verantwortungsvollen Quellen
Paper from responsible sources
FSC® C105338

If you have any concerns about our products,
you can contact us on
ProductSafety@springernature.com

In case Publisher is established outside the EU,
the EU authorized representative is:
Springer Nature Customer Service Center GmbH
Europaplatz 3, 69115 Heidelberg, Germany

Printed by Libri Plureos GmbH
in Hamburg, Germany